The Illustrated Guide to PSpice

The Illustrated Guide to PSpice

Robert Lamey

Cincinnati Technical College

DELMAR PUBLISHERS INC.™

I(T)P™

NOTICE TO THE READER

Publisher does not warrant or guarantee any of the products described herein or perform any independent analysis in connection with any of the product information contained herein. Publisher does not assume, and expressly disclaims, any obligation to obtain and include information other than that provided to it by the manufacturer.

The reader is expressly warned to consider and adopt all safety precautions that might be indicated by the activities described herein and to avoid all potential hazards. By following the instructions contained herein, the reader willingly assumes all risks in connection with such instructions.

The publisher makes no representations or warranties of any kind, including but not limited to, the warranties of fitness for particular purpose or merchantability, nor are any such representations implied with respect to the material set forth herein, and the publisher takes no responsibility with respect to such material. The publisher shall not be liable for any special, consequential or exemplary damages resulting, in whole or in part, from the readers' use of, or reliance upon, this material.

Cover design by Katie Hayden

Delmar Staff:

Administrative Editor: Wendy J. Welch
Senior Project Editor: Christopher Chien
Senior Production Supervisor: Larry Main
Art and Design Coordinator: Lisa Bower
Assistant Editor: Jenna Daniels

For information, address Delmar Publishers Inc.
3 Columbia Circle
Box 15-015
Albany, New York 12212-5015

Printed in the United States of America
Published simultaneously in Canada
by Nelson Canada,
a division of The Thompson Corporation

1 2 3 4 5 6 7 8 9 10 XXX 00 99 98 97 96 95 94

Library of Congress Cataloging-in-Publication Data

Lamey, Robert.
The illustrated guide to PSPICE / Robert Lamey.
p. cm.
Includes index.
ISBN 0-8273-6524-1
1. PSpice. 2. Electric circuit analysis—Data processing.
3. Electronic circuits—Data processing. I. Title.
TK454.L35 1995
621.3815'0285'5369—dc20

94-2662
CIP

Contents

Trademarks

"PSpice" is a registered trademark of MicroSim Corporation.

"The Design Center," "Probe," "Schematics," "StmEd," and "Stimulus Editor" are trademarks of MicroSim Corporation.

"Microsoft," "MS," and "MS-DOS" are registered trademarks, and "Windows" is a trademark of Microsoft Corporation.

"IBM" is a registered trademark of International Business Machines Corporation.

Preface

While SPICE software for computer simulation of analog circuitry has been used commercially for years, I was first introduced to PSpice (SPICE for personal computers) just two years ago. The impact on my teaching style and on my students was immediate and dramatic.

It also became clear that working technicians and electronics hobbyists could benefit greatly from PSpice. PSpice takes the tedium out of circuit analysis calculations, which can bog down a lecture and make discoveries in circuit performance boring rather than exciting. This is particularly true when recalculating the same circuit after making a change in the value of a single component or the frequency of a single source. PSpice, used as a demonstrator, added accuracy to the waveform phase shifts and frequency response curves that was marginal when graphed by hand on a blackboard.

For my students, the advantages were even greater. PSpice has become an integral part of every laboratory experiment. Simulation provides an accurate theoretical model that students use for comparison to the circuits in the lab exercise. Discrepancies are discussed in terms both of sources of experimental error and of necessary changes in the model to better explain the voltages and currents in the circuit under examination.

The introduction of PSpice into my analog circuit analysis classes prompted the search for an appropriate textbook. I quickly found that all the available books presume the reader is approaching PSpice with the background of an engineer, and is familiar with terms such as "transient analysis," "quadratic voltage coefficient," and "base-emitter zero bias p-n capacitance." The authors maintain a focus on the completeness that an engineer demands, at the expense of the directness and clarity that a student requires. The reference book that fills the engineer's needs is virtually unreadable to a technical student, or the amateur hobbyist, and sometimes even the experienced technician who is using a circuit simulator for the first time.

There seemed to be no textbook specifically written to explain how to use PSpice to model the circuits students typically encounter in a technical curriculum, or an amateur encounters in pursuit of his hobby. This book has been designed to fill the need of the student, the technician, the hobbyist, and the computer simulation novice.

To achieve this end:

1. Features that a student or hobbyist will likely use in the course of his technical education, or in the pursuit of certifications and licenses, are included. All other features are either relegated to the appendixes or omitted.
2. Material is presented in the order it is usually encountered in an electronics engineering technology curriculum. This chronology provides order for the student and easy reference for the amateur electronics maven. Following an introductor chapter about PSpice itself, Chapter 2 explains how PSpice can be used to analyze circuitry found in a DC circuit analysis class. Chapter 3 uses PSpice for AC circuit analysis. This curriculum based, chapter-wise progression continues through the study of diodes, op-amps, transistors, and communication circuitry.
3. All features presented are also illustrated with examples. In each example, new features are explained in sufficient detail to allow the reader to use the new feature in an assignment, a lab exercise, or to apply to circuitry under test for trouble shooting purposes.

This book is not complete and makes no attempt to cover every option to every PSpice command. Completeness is left to other authors and other books. Here the focus is on clarity, relevance for electronics students, technicians, and hobbyists, and immediate applicability to topics covered in an electronics engineering technology curriculum.

Although the author of this text is singular, many contributed their time and effort before the final product was complete. I would like to acknowledge the contributions of Roberto Uribe, Kent State University; Innocent Usoh, Nashville State Technical University; Ian Davis, Miami Dade Community College; Adnan Al-Smadi, Tennessee State University; and Ronald Harris, Weber State University, Utah. These readers scrutinized several chapters, suggesting improvements and detecting errors. Gary Webster and Robert McLain, my colleagues at Cincinnati Technical College, provided the sounding board against which my ideas for material and presentation were tested. My special thanks are reserved for Andrea Feld-Brockett, who spent many tedious days editing the original manuscript.

I am particularly indebted to the people at MicroSim Corporation for not only generously allowing portions of their software documentation to be reproduced in the appendixes of this book, but also for their permission to package the evaluation version of PSpice with the text.

Recognition should also be given to the students of Cincinnati Technical College, who patiently made use of this text in manuscript while it was still being shaped and reshaped into its final form.

1

Getting Started with PSpice

1.1 PSpice Software

In order to free engineers (and students) from the laborious, sometimes complex, and usually time-consuming work of circuit analysis, a team at the University of California at Berkeley developed SPICE (Simulation Program with Integrated Circuit Emphasis) software in the mid-1970s. SPICE consisted of a set of powerful algorithms for a wide range of circuit analysis methods. Unfortunately, SPICE was available only for mainframe and VAX-type computers.

In 1984 MicroSim Corporation made SPICE analysis available for personal computers under the name PSpice. Although PSpice is a commercial product costing thousands of dollars (depending on platform and analysis features), MicroSim has chosen to make an evaluation version available at no cost.

The evaluation version of PSpice has many, but not all, of the features of the commercial product. This version, while fully functional, has been designed with limitations that make it inadequate for commercial use. Businesses are expected to purchase the commercial PSpice package. For classroom and evaluation use, the limitations are not significant. The most relevant of these limitations are:

1. The evaluation version restricts circuits to a maximum of 10 transistors, compared to 200 transistors for the commercial package. This is not usually a problem in student designs. When necessary, the limitation can be circumvented by the creation of a subcircuit using a dependent current source to model the behavior of the transistor. There are similar limitations and solutions for the number of operational amplifiers that can be simulated.
2. The library for the evaluation version is much smaller than that of the commercial package. However, the reduced library does contain the more common components used in technical textbooks. For example, the 741 op amp, the 2N2222 transistor, and the 2N3906 and 2N3904 matched pair are included in the evaluation library.

There are other restrictions, such as the inability to perform distortion analysis; but this and similar restrictions are not important in a technical electronics program.

1.2 Installation

PSpice must be run from a hard disk. Evaluation Version 6.0 is the current version of the PSpice software as of this publication date, and has been supplied with this text. The evaluation version may be copied and shared without violating copyright laws. Users can also contact MicroSim Corp. to obtain evaluation copies free of charge. Since Delmar has bundled PSpice with this text, contacting MicroSim would only be necessary if the hardware of the user's system required a different version. (See Section 1.4—PSpice Hardware Requirements.)

MicroSim will supply PSpice on 3.5- or 5.25-inch floppy disks for a DOS, Windows, or a Macintosh environment. The following installation description assumes that PSpice is being loaded onto the C: drive from 3.5-inch floppies in the B: drive and will run under DOS. Installation under other conditions is very similar.

To install PSpice, place disk one, which contains the INSTALL program, in drive B: and change to B: drive. At the B: prompt, type:

```
B:\>install <return>
```

The INSTALL program is virtually transparent to the user. That is, with a minimum of user input, INSTALL will decompress the PSpice files, transfer the files into an appropriate subdirectory, and make changes to the user's AUTOEXEC.BAT and CONFIG.SYS files. The INSTALL interface to the user is a series of windows that open to explain the installation process, or to prompt the user for information. A detailed window-by-window explanation of the process is neither necessary or possible since INSTALL examines the user's computer and opens different windows depending on the system it finds. The process is well designed and painless; just answer questions, hit keys when prompted, and load the next disk when asked.

PSpice will be loaded into a subdirectory. The default name is \MSEVAL60. This is slightly cryptic (it stands for MicroSim Evaluation Version 6.0) and inconvenient, and the user is given the opportunity to use a different subdirectory name by backspacing over the old name and entering a new one. The author prefers \SPICE, and so \SPICE will be used throughout this book.

After a subdirectory name is chosen, INSTALL will decompress and transfer the PSpice files. When this is complete, INSTALL may request permission to alter the user's AUTOEXEC.BAT and/or CONFIG.SYS files. If INSTALL asks if it may change a file, give it permission; the changes are necessary. The PSpice software requires a path into \SPICE and the setting of an environmental variable (SIMLIBPATH = C:\SPICE) in AUTOEXEC.BAT. It also requires proper "file" and "buffer" sizes in the CONFIG.SYS. These details need not concern the user; INSTALL will set all parameters correctly.

The final messages to the user are to: (1) run the program SETUPDEV.EXE, which is necessary; (2) read the README.DOC file, which is not necessary; and (3) re-boot the system to use the new AUTOEXEC.BAT and CONFIG.SYS. Running SETUPDEV.EXE is important and cannot be skipped. PSpice needs to know a few facts about the user's system that INSTALL cannot determine; specifically, the type of display, the port to use for making hard copies, and the brand and model of the hard copy device. SETUPDEV is a menu-driven program and allows the user to choose a video driver (probably IBMVGA), hard copy port (probably LPT1), and hard copy device from lists of options. Be sure to choose the SAVE option in the main menu before exiting.

Reading README.DOC can be skipped. The information is of little use to the new user of PSpice. README.DOC contains upgrade information and enhancement features that are only useful to the professional designer. Re-booting the system is important. While the AUTOEXEC.BAT can be rerun from the DOS prompt, changes made to the CONFIG.SYS file are unavailable to the system until a re-boot occurs.

1.3 Working Directory

Since INSTALL has created a path into C:\SPICE, the PSpice files can be run from the root directory, but this leads to the clutter of many circuit files stored in the root. PSpice can be used from the \SPICE directory, but this places user files among the PSpice files and creates more clutter. Another option is to work from the \SPICE directory but name every circuit file with A: (or B:) so it is saved on a floppy. This requires the user to always have a floppy. It also makes save and retrieve operations slow because of the slow access times for floppy drives.

The best solution to this problem is to create a new directory just for the user's circuit files. This method is used in this book. To create the new directory, change into the \SPICE directory and type:

```
C:\SPICE>mkdir circuits
```

This will make a new subdirectory under \SPICE. When using the PSpice program, first change into this directory using the change directory command. Type:

```
C:\>CD \SPICE\CIRCUITS <return>
```

The path into \SPICE created by the INSTALL program makes the PSpice programs available while in the \SPICE\CIRCUITS subdirectory, but files created by the user will be isolated in their own subdirectory.

1.4 PSpice Hardware Requirements

The hardware requirements for PSpice have grown over the last several versions as the software has been improved in its analysis abilities and its interface. Version 6.0 requires a hard disk, a co-processor and at least 2 Megabytes of extended memory. Users with systems that do not meet these requirements must either upgrade their systems or download earlier versions of the software from the MicroSim BBS (714–830–1550; no parity, 8 bits, 1 stop bit). Version 5.4 required a co-processor, but could operate in conventional memory, i.e. 640 kbytes of RAM. Version 5.1 worked in conventional memory and could operate with or without a co-processor; however operation without a co-processor was considerably slower.

A mouse is optional in all versions. Versions 5.1 and 5.4 used a text-based editor that was usually more convenient to operate from the keyboard. Version 6.0 gives the user a considerably improved graphic editor that can operate from the keyboard but makes much better use of the mouse by providing pull-down menus and hot buttons. While the monitor is obviously necessary, a printer (or plotter) is not. PSpice will still operate even if there is no means for making a hard copy of its results.

1.5 PSpice Editor

The most apparent improvement in version 6.0 of PSpice is the editor. The old, slow, text-based control shell and editor have been replaced by a graphic windows-style editor that is intuitive, convenient, and responds better to a mouse. Since the purpose of this text is to provide information about PSpice on an as-needed basis, a complete description of the editor operation is not given here. Editor features, like the PSpice analysis features, are introduced as they become relevant through the text. For now, an overview description of the editor will suffice.

The program PS.EXE is used to enter the PSpice editor. To begin using PSpice, change into the \SPICE\CIRCUITS subdirectory and type

C:\SPICE\CIRCUITS>PS (filename)<return>

The user has the option to provide PSpice with the name of a file only if the file already exists. PSpice will not create a new file at the command line. When creating a new file, type PS with no argument. New files are named within the PSpice editor. When the return key is hit, PSpice displays the screen as shown in Figure 1.1.

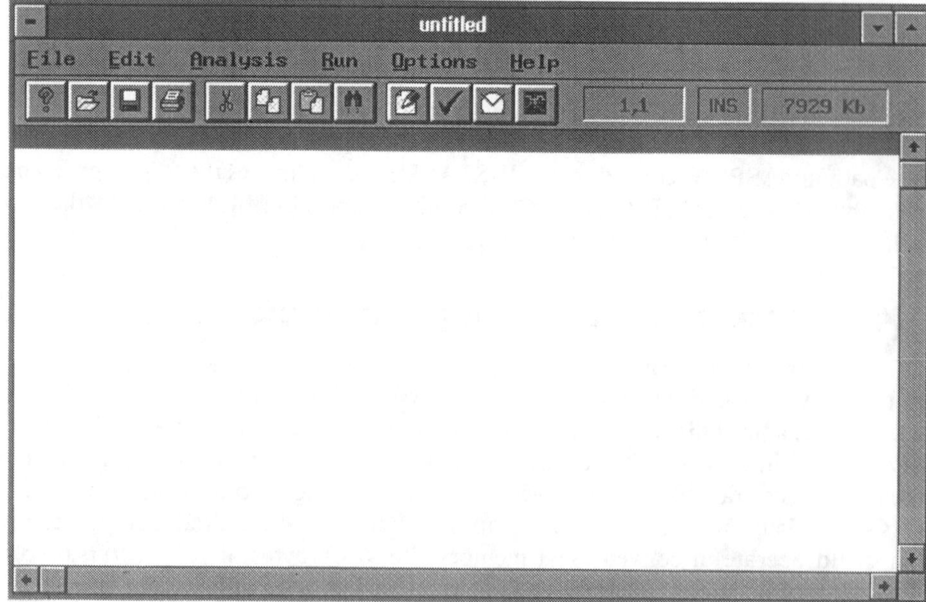

Figure 1.1

The top line of the window contains a minus sign (–) on the left. A double click with the left mouse button on the minus sign is the signal to exit the PSpice editor and return to the DOS prompt. In the center is the filename of the current circuit description. This remains "untitled" until the first save of the circuit file. To the right are two squares containing small triangles. A single click with the left mouse button on the first square shrinks the editor window to an icon. A double click on the icon is required to return the editor window. A single click on the second square expands the window to fill the entire screen.

NOTE: In most programs the left mouse button is used far more than the right button. Consequently, it has become common to omit the reference to left or right, and to assume that "to click the mouse" always refers to the left button. When the right button is to be used, the word "right" is explicitly indicated. This convention will be used for the rest of this text.

The second line in the window lists menu titles for six pull-down menus. These are accessed by clicking with the mouse or by holding down the <Alt> key and hitting the underscored letter in the menu title. Most of the menu items are obvious, making PSpice very intuitive to use. For example, most users would guess that pulling down the "File" menu and choosing "Save" would save the file to disk. Choosing "File" and then "Print" sends the file to the hard copy device (printer). Each menu option will be described as needed. Because saving files and opening already existing files are the primary interactions between PSpice and the operating system, a few items should be mentioned about naming files. DOS requires the root name of a file to be eight characters or less and PSpice requires a .CIR extension. When saving files, the extension can be included in the file name or omitted. The PSpice editor will automatically add the extension .CIR. Files should be named to indicate the type of circuit described in the file, but this is not mandatory. A filename of FOOLISH.CIR would be accepted, but this would be foolish. Files should be named to indicate a lab exercise or a problem number, or given names such as SERIES.CIR or RESONANT.CIR.

The third line in the window contains the hot buttons. Clicking a button is a shortcut to performing some of the most common operations. These are quite convenient and in most cases obvious. Clicking the question mark opens the help files. The second, third, and fourth buttons open, save, and print files respectively. The other buttons will be discussed as needed. The three rectangles to the right of the buttons contain the current cursor position, the editor mode (either insert or overwrite), and the amount of RAM memory in the computer.

Below these lines is the text area for writing the circuit description. To the right and below the text area are the scroll bars; these are accessed by the mouse, and allow the user to scroll through a long or wide document.

1.6 PSpice Circuit Descriptions

For PSpice to do its work, it needs information from the user about the circuit to be analyzed. This information consists of the following items.

1. THE TITLE LINE. The first line of every circuit description file is considered the title line and ignored by the analysis algorithms. It is not necessary to title a file, but it is necessary to provide a first line to be ignored. A typical title line might look like this:

   *Series circuit—Three resistors

   The asterisk (*) is recognized by PSpice as the beginning of a comment statement, and everything from the asterisk to the end of the line is ignored. Since the first line is always ignored, the asterisk is not strictly necessary; but it is generally used for clarity. If the user does not intend to title the circuit, a lone asterisk can be placed on line one.

   One of the most common mistakes made by students learning PSpice is forgetting that line one will be ignored. If a voltage source or electrical component is placed on line one, it will be ignored when the circuit is analyzed.
2. SOURCES. PSpice accepts both voltage and current sources. Sources can be direct current or a variety of alternating current waveforms, and multiple sources can be included in a single circuit. The naming of sources and the proper syntax for use in a circuit description is left for Chapter 2, where DC circuits are described and analyzed.
3. COMPONENTS. PSpice accepts both passive and active components. Passive components include resistors, capacitors, and inductors. Diodes, bipolar transistors, op amps, and FETS are active components; that is, they perform amplification or switching functions, or they require their own source of power to perform their function. Again, procedures for naming and using these components are left to appropriate chapters.
4. ANALYSIS. PSpice can analyze circuits in many ways. It can determine component voltage drops and branch currents, but it can also sweep frequencies to determine bandwidth or perform Fourier analysis on complex waveforms.
5. OUTPUT. Once the circuit has been analyzed, the information requested by the user can be displayed as specific voltages and currents in an output file. The output file has the same root name as the circuit description file but has .OUT instead of .CIR for the extension. PSpice can also display the output information graphically on the monitor. By using PROBE to display output, the user can turn the computer screen into an oscilloscope and view waveforms as a function of time. PROBE can also display the results of a Fourier analysis effectively simulating a spectrum analyzer. If the circuit description file contains an error, the location of the error and the type of error are noted for the user in the output file.

1.7 Values and Units in PSpice

Values in electrical circuits vary greatly in magnitude. A circuit may have a 3 picoFarad capacitor and a 22 megohm resistor. PSpice allows users to enter component, source, and analysis values in standard floating-point notation, exponential notation, or by using prefixes. For example, a 3 microHenry inductor can be expressed as

.000003 or 3e-6 or 3u

First notice that the unit Henry is not necessary. The user can include the proper unit for clarity, but PSpice ignores words such as Farads, Ohms, and Seconds. That is,

.000003Henry or 3e-6Henry or 3uH

are also valid inputs. PSpice knows the proper units for capacitors, resistors, inductors, time, and the like. Consequently the 22 megohm resistor can be expressed equivalently as

22000000	22000000Ohm	22000000Ω	22Meg
22Megohm	22e6	22e6Ohm	2.2e7
22000K	22000Kilo	22000KΩ	22000e3

Capacitor values deserve extra caution. If the word *Farad* is used following a capacitor value, the initial *F* in *Farad* is interpreted by PSpice as the prefix *femto*. The following two expressions do not represent the same capacitor value.

3e-6 3e-6Farad

The first expression indicates 3 microFarads. The second value is 3e-6 femtoFarads or 3e-21 Farads. To include the units and retain the value of 3 microFarads, the prefix for "micro" should be used instead of the exponential notation. The following two expressions are equivalent.

3e-6 3uFarad

Valid prefixes recognized by PSpice are listed here:

F	(FEMTO)	$10e^{-15}$
P	(PICO)	$10e^{-12}$
N	(NANO)	$10e^{-9}$
U	(MICRO)	$10e^{-6}$
M	(MILLI)	$10e^{-3}$
K	(KILO)	$10e^{+3}$
MEG	(MEGA)	$10e^{+6}$
G	(GIGA)	$10e^{+9}$
T	(TERA)	$10e^{+12}$

These prefixes can be used as upper- or lower-case letters. PSpice is not case-sensitive, meaning upper- and lower-case letters are treated as identical characters. The use of most of these prefixes is obvious, but special note should be made that the symbols for micro, milli, and mega are U, M, and MEG.

1.8 PSpice from the Command Line

All PSpice operations such as printing, analysis, and graphical display can be accessed from within the PSpice editor, making the editor serve as the controller for the PSpice environment. This makes PSpice convenient and accessible to users who are only marginally familiar with DOS or perfer the convenience of a self-contained environment. The user does have the option to use PSpice from the command line, using his or her own preferred editor (any editor that will create an ASCII file), and calling the analysis and display programs directly from the DOS prompt. This was a useful feature in earlier versions of PSpice because the editor was slow and responded poorly to the mouse. In version 6.0, the editor has been greatly improved, making command line operation a matter of personal preference.

In this text, PSpice features will first be explained using the PSpice editor and afterward using command line entries. This is primarily for users who, because of hardware limitations, are using earlier versions of PSpice. DOS 5.0 and DOS 6.0 include a text editor called EDIT. Since this editor is as common to DOS users as any, EDIT will be used for examples of using PSpice from the command line.

2

DC Circuits

The focus of Chapter 2 is the description and analysis of circuits containing one or more direct-current power sources. The circuits to be analyzed are circuits that are likely to be encountered by students in a DC circuits analysis class or by novice electronics hobbyists. Circuit complexity increases through the chapter from a simple series circuit; through multiloop, multisource circuits; to DC circuits using capacitors and inductors. The description and analysis of Example 2.1 (a single-source, three-resistor series circuit) are explained in minute detail, including the use of the PSpice editor and the simulation software. For subsequent circuits, only new features are given this level of scrutiny.

2.1 Circuit Description

The circuit shown in Example 2.1 must be described in a manner understandable to PSpice, whether the PSpice editor or another editor is used. We use the PSpice editor first and discuss alternatives later.

For reasons described in Chapter 1, it is assumed that the PSpice files are in a subdirectory called \SPICE and the working directory is a subdirectory under \SPICE, namely \SPICE\CIRCUITS. At the appropriate prompt, type PS followed by <return>.

```
C:\SPICE\CIRCUITS>PS <return>
```

This command will call the PSpice environment. If this is the first time that PSpice has been used, or if a new blank text area was left in the editor when PSpice was last

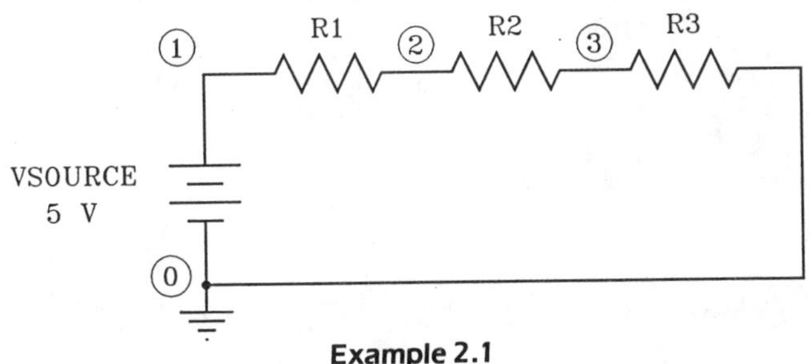

Example 2.1

exited, "untitled" appears in the center of the top line. If a circuit description was left in the editor, PSpice remembers the last circuit description and loads it. Figure 2.1 shows the menu options available in the "File" pull-down menu. Choosing "New" from the menu replaces an existing circuit description with a blank screen, ready for a new circuit.

In order to be both concise and clear about references to menu items and to actions requiring a mouse click, two typesetting conventions are adopted for the editor. First, references to menus, menu selections, or certain pop-up windows that appear in the editor are indicated by the use of *italics*. For example, the previous paragraph refers to the *New* selection in the *File* pull-down menu. Second, actions requiring a mouse click are indicated by **boldface** type. Sequences of actions are separated by backslashes. Saving a file to disk requires clicking the *File* menu and selecting *Save*. These actions are indicated by **\File\Save.** In fact, menu selections can be made in three ways:

1. From the keyboard, pull-down menus are chosen by hitting the <Alt> key and the underscored letter in the menu title. Menu items are chosen by hitting the underscored letter without the <Alt> key.
2. From the keyboard, pull-down menus can be chosen by hitting the <Alt> key, using the cursor keys to move the highlight left or right, and hitting <return>. When the menu opens, move up or down using the cursor keys to the required selection and hit <return> again.
3. Most commonly, menu selections are made using the mouse, clicking on each indicated **boldfaced** item.

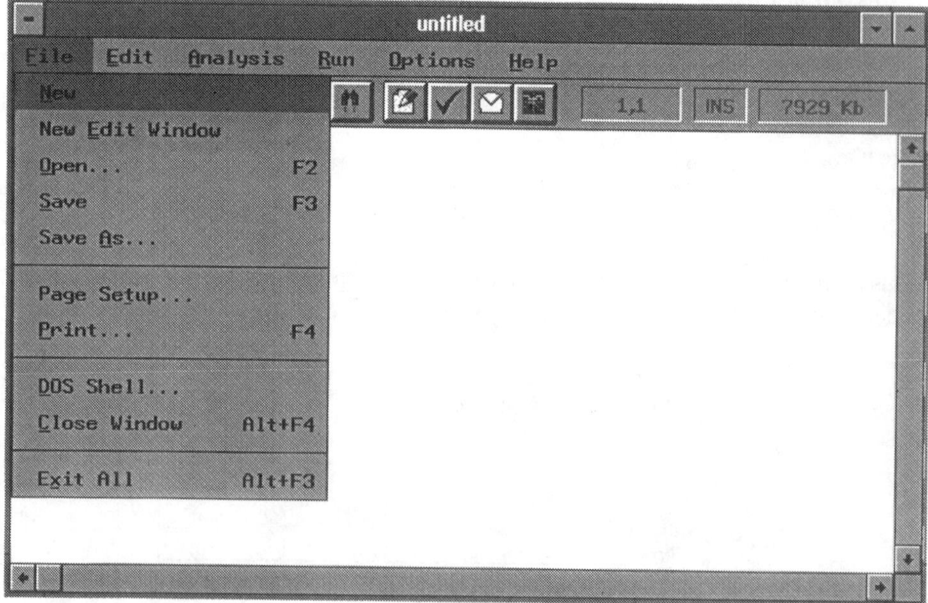

Figure 2.1

With the editor open and a blank text area (using **\Files\New** if necessary), the circuit description can be entered. The circuit description is a list, telling PSpice about the circuit to be analyzed. Since computers are notoriously poor at reading users' minds, care must be taken to describe the circuit in terms that PSpice can understand. The circuit description for Example 2.1 is given below with explanation for the proper syntax.

Title Line The first line of the circuit description—the title line—is ignored by the analysis program. Begin the title line with an asterisk. PSpice is not case-sensitive, meaning upper- and lower-case letters are treated identically. Throughout this book, upper-case letters are used for uniformity.

*DC1.CIR – SERIES CIRCUIT

Source Description PSpice allows the use of voltage and current sources. The general description of a DC voltage or current source is

<source> <pos node> <neg node> <DC value>

where <pos node> and <neg node> are the positive and negative insertion points, respectively, for the DC supply. The circuit description becomes

*DC1.CIR – SERIES CIRCUIT
VSOURCE 1 0 5

There are several items to note in the description of a DC source.

1. A voltage source begins with a "V" plus the source name. A current source begins with an "I" plus the source name. The source name is any string of alphanumeric characters. Examples of acceptable source names are VSOURCE, VDC, V1, V38A, and IALPHA2. No space is permitted between the V (or I) and the rest of the source name. PSpice uses white space for delimiters; and entering V SOURCE would result in a double error because V would have no name and SOURCE would be considered the name of <pos node>.
2. The order of the nodes is important. The positive node is written before the negative node. If the order is reversed, currents will travel through the circuit in an unexpected direction. Any alphanumeric string (up to 131 characters) can be used to name a node, but in practice nodes are usually named using non-negative integers.
3. PSpice uses conventional current flow. That is, positive current moves from positive to negative through a load. Since conventional current flows from negative to positive through a voltage source, this current is given a negative value by PSpice.
4. The circuit must have a ground reference named node "0" (node zero). Often it is convenient for the ground to be at the negative end of the source, but it can be at any point in the circuit.

5. PSpice understands that the "5" means 5 Volts. The word "Volts" can be included to make the circuit description more readable for the user, but it is not necessary. If the word "Volts" is included, no space is permitted between the numeric value and the word "Volts," i.e., 5Volts.
6. The general description for any voltage or current source (DC or any of several AC waveforms) is given in Appendix C.
7. The delimiters between VSOURCE, the nodes, and the voltage value are spaces. PSpice also accepts multiple spaces, tabs, and commas. Throughout this text, spaces are used.

Component Description In this circuit, only resistors are used. In PSpice, resistors are indicated by an "R" plus a name. A complete list of component symbols is given in Appendix C. For the sake of simplicity, the resistors are called R1, R2, and R3. In more complicated circuits, components are usually given more descriptive names to make the description more readable. For example: RLOAD, REMITTER, RCOIL, or RSOURCE.

The general form for describing a resistor is

R<name> <pos node> <neg node> <resistor value>

```
*DC1 CIR – SERIES CIRCUIT
VSOURCE 1 0 5
R1          1 2 1K
R2          2 3 2K
R3          3 0 3K
```

Just as in the case of the source description, the nodes of components are listed positive node first.

Analysis In a resistive DC circuit, the user is generally interested in component voltages, voltages at nodes relative to ground, and branch currents. The general form of the command to perform .DC analysis is

.DC <source> <starting voltage> <ending voltage> <increment>

Several items require further explanation in the .DC analysis command.

1. The dot (period) preceding the DC is required and indicates a directive for the PSpice software, as opposed to the previous component descriptions.
2. .DC analysis uses a "sweep" of voltage levels for a particular source. The first argument following the .DC command is the voltage source that is to be varied over the range set by the next two values. In this example VSOURCE is the source; but the circuit to be simulated does not have a variable source.
3. In order to indicate that a single voltage level is to be used, the starting and ending voltages are given the same value. An increment value must be provided,

even though the value is irrelevant, since the beginning and end of the sweep are the same voltage. PSpice requires an increment value greater than zero. For the example here, an increment of 1 is chosen arbitrarily.

The circuit description for Example 2.1 becomes

```
*DC1.CIR—SERIES CIRCUIT
VSOURCE 1 0 5
R1        1 2 1K
R2        2 3 2K
R3        3 0 3K
.DC       VSOURCE 5 5 1
```

Output PSpice can display the results of the analysis either graphically or as a printed list of values. In this example, because the values of concern are static and do not change over time, making a list is more appropriate. PSpice will deposit these values in a file with the same root name as the circuit description file, but with an .OUT extension, i.e., DC1.OUT. The output file will be examined in detail when the circuit description is complete and the PSpice analysis is finished. The general form for the output display command is

.PRINT DC <value one> <value two> . . .

The .PRINT DC command will print any number of circuit voltages and currents. Component voltages are indicated by using the nodes surrounding the component or by using the component name. The voltage dropped by R2 is written

V(R2)　　　or　　　V(2,3)

Voltages at nodes relative to ground are written

V(2)　　　and　　　V(3)

for the voltages at node 2 and node 3 respectively.

Currents must be written using component names. Using nodes will create a syntax error. In a series circuit there is only one value for current, but in the .PRINT DC statement it can be written using any of the following.

I(R1)　　　I(R2)　　　I(R3)　　　I(VSOURCE)

PSpice uses conventional current, meaning current flowing from a positive node to a negative node through a load is considered positive. In the sample circuit, the current through R1, R2, and R3 travels from the positive end of each resistor to the negative end. The current through the source flows from the negative end to the positive end. Consequently, the value given by PSpice for I(SOURCE) will be the same magnitude as the current value for the resistors, but will have a negative sign.

Assuming the user is interested in the voltage dropped by R2, the voltage at node 2 relative to ground, and the current in the loop, the circuit description becomes

```
*DC1.CIR–SERIES CIRCUIT
VSOURCE 1 0 5
R1          1 2 1K
R2          2 3 2K
R3          3 0 3K
.DC         VSOURCE 5 5 1
.PRINT DC   V(2,3) V(2) I(R2)
.END
```

All circuit descriptions end with the .END statement.

This completes the circuit description. To save the circuit description to disk, either use **\File\Save** or click the third hot button from the left, i.e., the button with the picture of a floppy disk. If the file has not yet been named, a window will open providing the opportunity to name the file. Any name may be chosen, but in the context of this chapter, DC1.CIR is a logical choice. The user may include the .CIR extension or allow the editor to automatically add it. After the save operation, the word "untitled" at the top of the window is replaced by the new file name.

2.2 Circuit Analysis

Before running the analysis program, certain analysis options should be set. Open the *Analysis Options* window by executing **\Analysis\Analysis Options.** The options to automatically run PROBE after analysis and to browse the output file after the analysis should be turned off. These options are toggled on and off by clicking the box to the left of the option. PROBE is the graphics display program. Since the circuit description has requested two voltage values and one current value, there is no data to represent graphically. The browser allows the user to examine the analysis output, but does not allow any editing of the file. Usually the output file contains some unneeded information, a number of page breaks, and a great deal of white space, that can be deleted. *PSpice* should be chosen as the simulator and the window can be closed by clicking **OK.**

The analysis can be started in three ways. **\Run\Simulator;** hitting the F11 key; or clicking the analysis hot button (the 11th button from the left) begins the simulation. The simulator program begins by replacing the editor screen with a display containing information about the analysis. For a circuit as simple as DC1.CIR, the analysis only takes a second, and then the editor returns to the screen. The output file created by the simulator can be examined by using **\Analysis\Browse Output.**

If editing of the output file is necessary, it can be edited by using the PSpice editor. Use **\File\New Edit Window** and then **\File\Open** to open DC1.OUT. Typically, there is a large amount of unneeded information, white space, and page breaks to be deleted. Because the PSpice editor scrolls rather slowly, it may not be the best choice for editing the output file. If the analysis is to be included in a larger file such as a lab report, it is usually more convenient to load the entire DC1.OUT file into a word processor and then edit.

VSOURCE	V(2,3)	V(2)	I(R2)
5.000E+00	1.667E+00	4.167E+00	8.333E−04

The requested voltages and currents are displayed in exponential notation and in standard units, i.e., Volts and Amperes. The voltage from node 2 to node 3 is 1.667 Volts. The voltage from node 2 to ground is 4.167 Volts, and the current through resistor R2 is .8333 milliamps.

2.3 Command Line Circuit Description and Analysis

The circuit description file can be created and the PSpice analysis completed without the use of the PSpice command shell. Any text editor or word processor that will create an ASCII file can be used to create the DC1.CIR file. Since EDIT.EXE supplied with DOS 5.0 is probably the most commonly available text editor, it is used as the example editor in this book. Most users choose to store the utility files supplied with DOS in a subdirectory called DOS. Consequently it is necessary to have a PATH into C:\DOS as part of the PATH command in the AUTOEXEC.BAT file. If the user prefers to use another editor, a PATH to the appropriate subdirectory is necessary. Assuming the user has installed the PSpice files in the \SPICE subdirectory and has created another subdirectory \SPICE\CIRCUITS as explained in Chapter 1, the circuit description file is created by typing

```
C:\SPICE\CIRCUITS>EDIT DC1.CIR <return>
```

Even though the PSpice editor has been greatly improved, the EDIT editor has several advantages. EDIT scrolls long documents better, responds better to the mouse, and is more convenient for deleting large amounts of text. Use of the editor is not explained here. It is left to the user to enter the circuit description, repeated here for convenience:

```
*DC1.CIR–SERIES CIRCUIT
VSOURCE 1 0 5
R1        1 2 1K
R2        2 3 2K
R3        3 0 3K
.DC       VSOURCE 5 5 1
.PRINT DC  V(2,3) V(2) I(R2)
.END
```

Save the file and return to the DOS prompt. To run the circuit analysis software, enter

C:\SPICE\CIRCUITS>PSPICE DC1.CIR <return>

The result of the analysis is stored in DC1.OUT. To view the output, to edit it, or to print a hard copy, use the editor with DC1.OUT as the argument.

C:\SPICE\CIRCUITS>EDIT DC1.OUT <return>

2.4 Errors

It is always possible that a circuit description contains an error. For example, R3 might be described as spanning nodes 3 and 4. This would indicate an open circuit that would be impossible for PSpice to analyze.

```
R3          3 4 3K
```

Omitting the final "1" from the .DC analysis line would be an error.

```
.DC         VSOURCE 5 5
```

Of course, if the user incorrectly enters R3 as a 4K resistor instead of a 3K resistor, PSpice—having no way of knowing this was incorrect—would analyze the circuit using 4K as the value of R3.

The PSpice environment has two methods for detecting errors in a circuit description. The first method involves running the syntax checker before performing the simulation. This can be done in any one of three ways. **Analysis\Syntax Check;** hitting the F10 key; or clicking the hot button containing the check mark (the 10th button from the left) all run the syntax checker. If there is no error in the circuit description, a pop-up window notifies the user. If an error is found, a different window opens explaining the location and the type of error. Clicking *Goto Error* returns the user to the editor on the line where the error was found.

Alternatively, the simulator can be used without checking for errors. If the analysis program detects an error, the analysis is stopped and a message appears in the analysis window telling the user to examine the .OUT file. The location and the type of error is noted in the circuit description in the output file. The error must be corrected in the .CIR file; corrections made in the .OUT file will not correct the problem.

2.5 Variable-Source Circuits

The tedium of circuit calculations performed by hand doubles and triples when the same circuit must be analyzed a second and third time because a single component is changed. If the voltage source used in Example 2.1 is replaced by a variable supply, the component voltages and the current must be recalculated for each new source voltage value. In Example 2.2, the supply is changed from 5 Volts to 10 Volts in steps of .5 Volts. This means the circuit must be analyzed 11 times, a task better suited to computers than to humans.

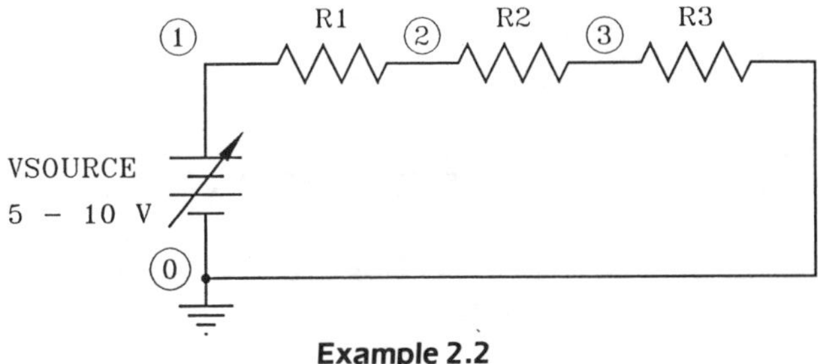

Example 2.2

To have PSpice sweep the voltages from 5 to 10 volts, the .DC analysis line is changed to read

.DC VSOURCE 5 10 .5

This instructs the software to analyze the circuit for all voltages from 5 to 10 Volts inclusively in .5 Volt increments. The output file will contain the following information:

VSOURCE	V(2,3)	V(2)	I(R2)
5.000E+00	1.667E+00	4.167E+00	8.333E−04
5.500E+00	1.833E+00	4.583E+00	9.167E−04
6.000E+00	2.000E+00	5.000E+00	1.000E−03
6.500E+00	2.167E+00	5.417E+00	1.083E−03
7.000E+00	2.333E+00	5.833E+00	1.167E−03
7.500E+00	2.500E+00	6.250E+00	1.250E−03
8.000E+00	2.667E+00	6.667E+00	1.333E−03
8.500E+00	2.833E+00	7.083E+00	1.417E−03
9.000E+00	3.000E+00	7.500E+00	1.500E−03
9.500E+00	3.167E+00	7.917E+00	1.583E−03
1.000E+01	3.333E+00	8.333E+00	1.667E−03

The task of recalculation changes from tedious to ridiculous when the voltage increment is decreased to something very small and requires hundreds or thousands of calculations. Even if the calculations are made by a computer, the resulting tabled data are difficult to read. A better and faster solution is to calculate fewer points and then graph those points, allowing the user to extrapolate the values in between the calculated values. PSpice can very simply make the indicated graph using the range of source voltages as the independent variable (X-axis), and the calculated component voltages and currents as dependent variables (Y-axis). To indicate that a graph is needed, the .PROBE command must be added to the circuit description. The revised file becomes:

```
*DC2.CIR
VSOURCE 1 0 5
```

```
R1          1 2 1K
R2          2 3 2K
R3          3 0 3K
.DC VSOURCE 5 10 .5
*           DC ANALYSIS FROM 5 TO 10 VOLTS - STEP .5 VOLTS
.PROBE
*           PROBE–COMMAND TO USE GRAPHICS TOOLS
.PRINT DC   V(2,3) V(2) I(R2)
.END
```

PSpice will still write the requested voltages and current to the .OUT file, but will also create a few new files that contain the information needed to graphically represent the results. PROBE.EXE is the PSpice program that accesses these files and displays the data on a coordinate plane. PROBE can be run automatically when the simulation is complete by clicking the *Auto-run Probe* box in the *Analysis Options* window **(\Analysis\Analysis Options),** or manually, using any one of three methods: **\Run\Probe;** hitting the F12 key; or clicking the PROBE hot button (the 12th button from the left).

As shown in Figure 2.2, PSpice allows the user to graph many voltages and currents. Generally, the source voltage is graphed first as a reference. A second and

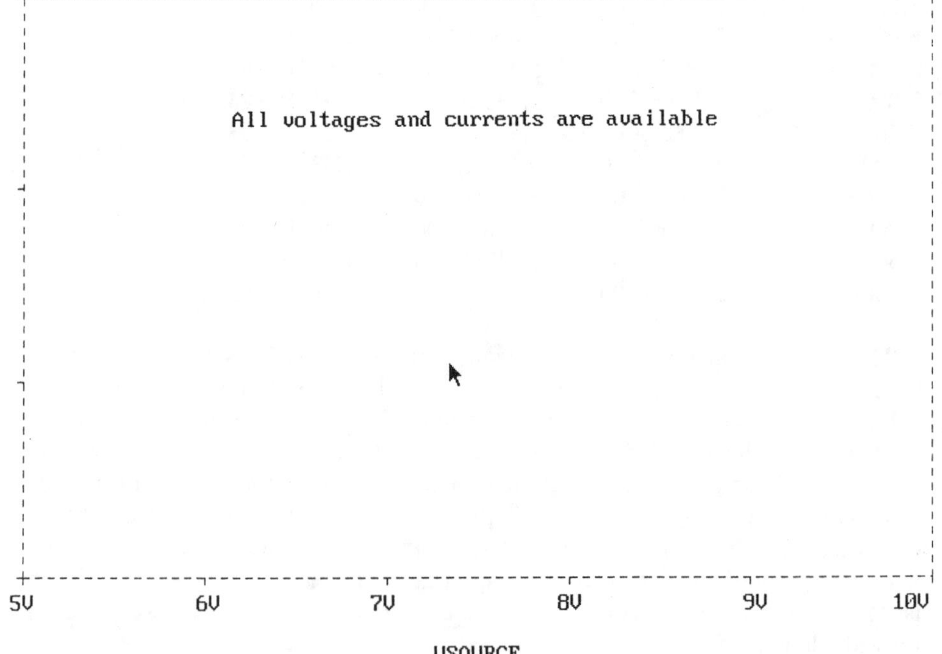

Figure 2.2

third trace can be added to show V(2,3) and V(2) as these voltages change with the changing source voltage.

The current I(R2) could also be graphed, but the plotted line would merely be traced across the bottom of the graph. This occurs because the current magnitude is about 1000 times smaller than that of the graphed voltages. PSpice is willing to use the Y-axis for both voltage and current, but the scale must be the same for both. This means that as the voltage at node 1 is graphed at 5 (5 Volts) the current at that voltage is graphed at .8333e-3 (.0008333 Amps) or virtually zero. The solution is to graph the current on its own plot so that PSpice can give it an appropriate scale on the Y-axis. To do this, choose "Plot_control" from the menu. This results in a new list of menu options. If you choose "Add_plot" from the menu, a second plot appears on the screen above the first. When you choose "Exit" from the "Plot_control" menu, the main menu reappears. The current can now be graphed with reasonable Y-axis scaling, as shown in Figure 2.3. Hitting "e" twice (or <esc> twice) returns the user to the PSpice editor. Working from the command line, the operation is virtually the same.

C:\SPICE\CIRCUITS>PSPICE DC2.CIR

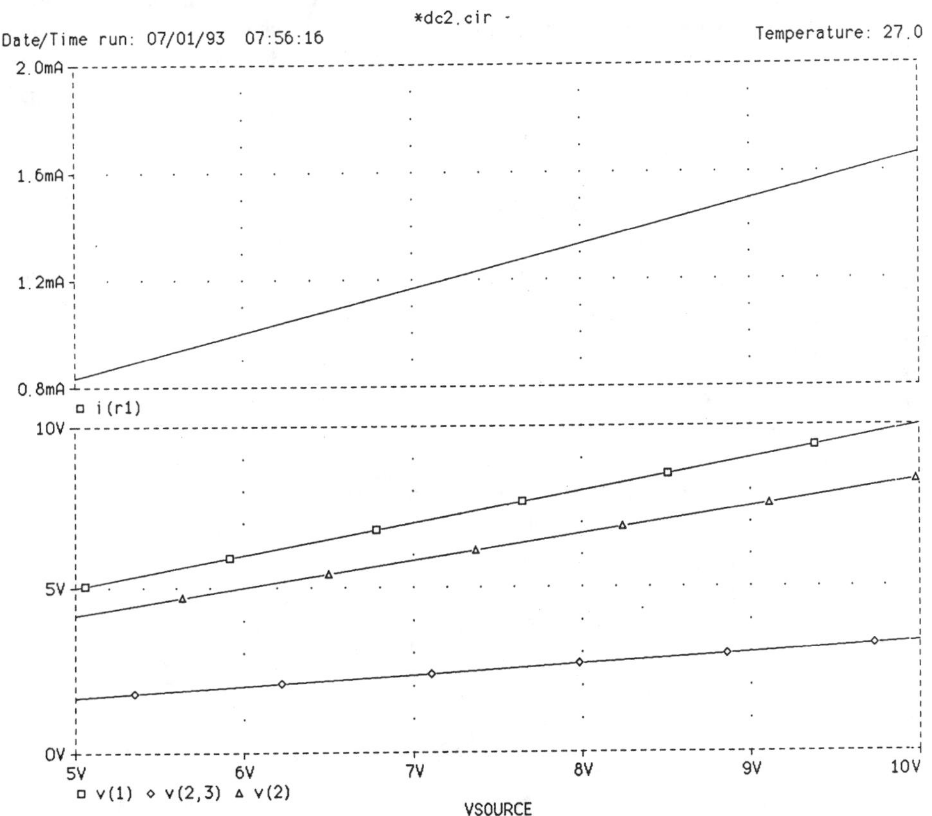

Figure 2.3

runs the analysis program. When the analysis is finished, PROBE can be run from the command line by typing

C:\SPICE\CIRCUITS>PROBE DC2.CIR

This two-command sequence has been automated in a batch file called SIM.BAT. By typing SIM DC2.CIR at the C:\SPICE\CIRCUITS> prompt on the command line, the computer is instructed to run the PSpice simulation program, and if the circuit description contains the .PROBE instruction, then the PROBE program is run automatically when the analysis is complete.

In fact, PROBE is more commonly used to illustrate changes over time or frequency than a range of voltages. Consequently, it is more often used to display voltages and currents in circuits with alternating current sources. Used in this manner, PSpice can duplicate the operation of an oscilloscope or a spectrum analyzer, but these functions will be explained in later chapters.

2.6 Series-Parallel Circuits

As the number of branches and elements in a circuit becomes greater, analysis by hand calculation becomes more difficult, more time consuming, and more prone to error. Each of these problems is minimized by using PSpice. Additional branches and elements require more care in labeling nodes and writing the circuit description, but otherwise the procedure for complex circuits is the same as that for a simple series circuit. See Example 2.3.

```
*DC3.CIR—SERIES - PARALLEL CIRCUIT
V1 1 0 20
R1 1 2 100
R2 2 3 200
R3 3 4 300
R4 2 5 400
R5 3 6 500
R6 5 0 600
```

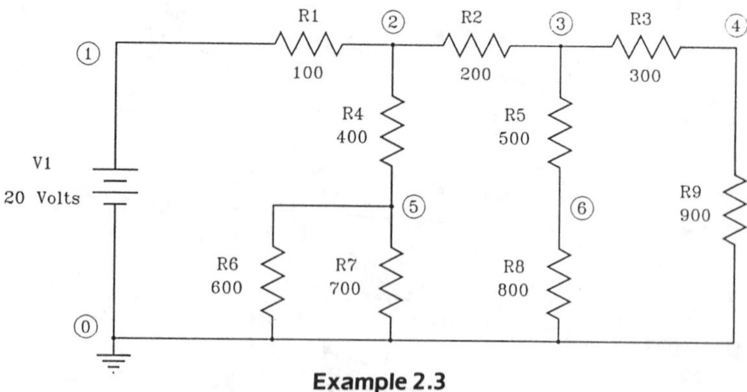

Example 2.3

```
R7 5 0 700
R8 6 0 800
R9 4 0 900
.DC V1 20 20 1
.PRINT DC I(R1) I(R4) I(R5) I(R9)
.PRINT DC V(2,3) V(3,6) V(6,0) V(0,5) V(5,2)
.END
```

The currents in the .PRINT DC statement in Example 2.3 were chosen to illustrate that the current provided by the source is equal to the sum of the branch currents. The current I(V1) is negative because the circuit current moving through the source is flowing from the negative end of the source to the positive end.

V1	I(V1)	I(R4)	I(R5)	I(R9)
2.000E+01	−4.123E−02	2.196E−02	9.249E−03	1.002E−02

The voltages in the second .PRINT DC statement were chosen to illustrate that the voltage drops and rises around a loop must sum to zero. The polarity of the requested voltages is particularly important when summing a loop. As the loop is traced in the clockwise direction, the values are positive until V(0,5). The voltage from node 0 to node 5 is a rise in potential and is represented by PSpice as a negative drop.

V1	V(2,3)	V(3,6)	V(6,0)	V(0,5)	V(5,2)
2.000E+01	3.854E+00	4.624E+00	7.399E+00	−7.094E+00	−8.783E+00

2.7 Multiple-Source–Multiple-Loop Circuits

The next level of difficulty usually encountered in the study of DC networks is the analysis of multiple-source–multiple-loop circuits. This involves solving simultaneous equations that describe each loop in the circuit, a process that greatly increases both the amount of time involved and the opportunity for error. Further, if the circuit analysis does not match the student's experimental results, tracing the error can be quite difficult. As circuitry increases in complexity, the advantages of using PSpice become more dramatic.

As in the case of series-parallel circuits, there are no new features to learn for the analysis of multiple-source–multiple-loop circuits. Errors in these circuits usually occur because of problems in describing the polarity of the voltage sources. Obviously, if a source is accidentally described to PSpice with the polarity reversed, the voltages and currents in the circuit will be changed radically.

Consider the circuit shown in Example 2.4.

```
*DC4.CIR   MULTI-LOOP, MULTI-SOURCE CIRCUIT
V1 1 0 5
V2 6 7 8
```

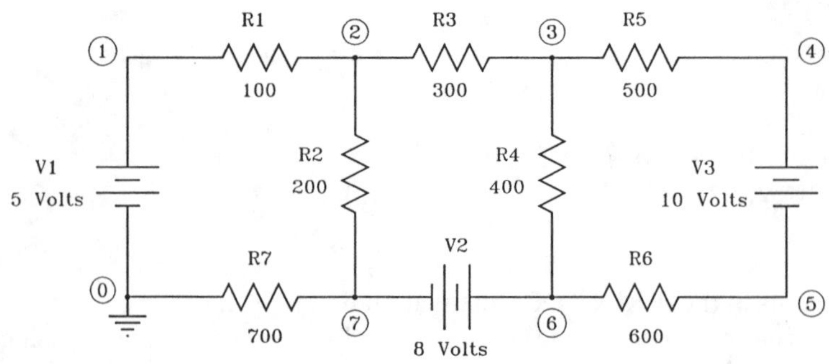

Example 2.4

```
V3 4 5 10
R1 1 2 100
R2 2 7 200
R3 2 3 300
R4 3 6 400
R5 3 4 500
R6 5 6 600
R7 7 0 700
.DC V1 5 5 1
.PRINT DC I(R4)
.END
```

The analysis of this circuit results in an output file containing the requested information.

```
*DC4.CIR  MULTI-LOOP, MULTI-SOURCE CIRCUIT
****      DC TRANSFER CURVES      TEMPERATURE =    27.000 DEG C

 V1              I(R4)
5.000E+00    -2.743E-03
```

In more simple circuits, it is usually easy to determine which end of a resistor is the positive node, and consequently should be placed first in the circuit description. In a circuit with multiple sources, the positive end of the component is less obvious. The current through resistor R4 was chosen for display in this example because it is not apparent in which direction current will flow through the resistor. In the circuit description, node 5 was placed first on the incorrect supposition that it was the positive node. PSpice is able to analyze the circuit, correctly giving the current through R4 a negative value to indicate that the current, in fact, travels backward compared to the polarity indicated by the user in the circuit description.

2.8 Analysis at Circuit Nodes

In all versions of PSpice prior to version 6.0, the .OUT for every circuit analyzed contained the voltages at each circuit node and the currents through each source. This served as the default analysis and required no special instruction in the circuit description. Version 6.0 does not provide node voltages unless specifically told to include them in the output file. The circuit description for the circuit in Example 2.4 has been rewritten with the .DC analysis and the .PRINT DC commands marked as comments by asterisks. .SAVEBIAS is the command to provide node voltages and source currents. The syntax for the command is

.SAVEBIAS <name of output file without .OUT extension><voltage type>

```
*DC4.CIR MULTI-LOOP, MULTI-SOURCE CIRCUIT
V1 1 0 5
V2 6 7 8
V3 4 5 10
R1 1 2 100
R2 2 7 200
R3 2 3 300
R4 3 6 400
R5 3 4 500
R6 5 6 600
R7 7 0 700
*.DC V1 5 5 1
*.PRINT DC I(R4)
.SAVEBIAS DC4 DC
.END
```

When the simulation is complete, the .OUT file contains the following information.

```
****   SMALL SIGNAL BIAS SOLUTION   TEMPERATURE =   27.000 DEG C

NODE VOLTAGE   NODE VOLTAGE   NODE VOLTAGE   NODE VOLTAGE

(   1)   5.0000  (   2)   4.7566  (   3)   8.6062  (   4)   13.6500
(   5)   3.6504  (   6)   9.7035  (   7)   1.7035

VOLTAGE SOURCE CURRENTS
NAME              CURRENT

V1                -2.434E-03
V2                -1.283E-02
V3                -1.009E-02
```

The node voltages are all referenced to node 0 (the ground point). The currents through the voltage supplies are negative because the current travels through supplies from negative to positive and PSpice indicates positive current as flowing from positive to negative. PSpice also allows the user to request the original information about the current through the R4 resistor by removing the comment asterisks from the .DC and .PRINT DC lines. If both types of analysis are requested, however, PSpice places the .DC analysis in the .OUT file and the analysis at the circuit nodes in a file without extension (DC4).

2.9 Transient Analysis of Resistor-Capacitor Circuits

The last topic usually considered in a DC circuit analysis class is the response of reactive elements to a DC supply. These circuits differ from all the previous circuits in that the currents and voltages generated are time dependent. That is, the current through the circuit and the voltages dropped across each component are not constant until the electric and magnetic fields of the reactive components have reached steady-state conditions. Consider the resistor-capacitor circuit shown in Example 2.5.

Example 2.5A is a circuit typically found in a lab exercise for determining RC time constants. When the switch is thrown to position A, current flows and the capacitor begins to charge. After the capacitor has reached full charge, the circuit is singularly uninteresting. After this point, no current flows, and the capacitor drops the entire source voltage. Consequently, analysis generally focuses on the action of the circuit prior to the full charge time. Similarly, in the discharge phase when the switch is moved to position B, the circuit is interesting only during the capacitor discharge. Afterward no current flows, and no voltages are generated.

The closing and opening of the switch in Example 2.5A can easily be described to PSpice by the use of a square wave source that outputs 5 Volts for some length of time and then changes to 0 Volts, allowing the capacitor to discharge. The PSpice circuit description for this circuit (Example 2.5B) introduces a few new features, so the description is given, and then the new items are explained.

```
*DC5.CIR   RESISTOR-CAPACITOR CIRCUIT
V1 1 0 PULSE( 0 5 0 1N 1N 5M 10M)
R1 1 2 1K
C1 2 0 1U
.TRAN 1M 10M
.PROBE
.PRINT TRAN V(2) V(1,2) I(R1)
.END
```

First, notice that the source is not described as a DC voltage source. The DC supply in PSpice is not time dependent. If the source were described as a DC supply, the output developed by PSpice would reflect the conditions after steady-state conditions had been reached. Besides this, the circuit requires a means of allowing the capacitor to discharge, simulating the switch being moved to position B.

Describing the source as a square wave that changes from 5 Volts to 0 Volts after

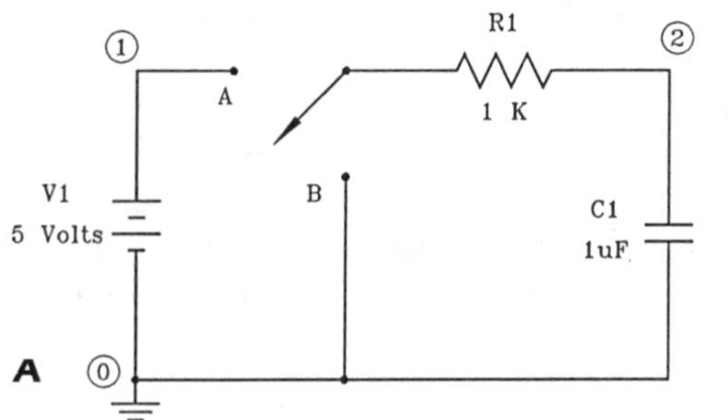

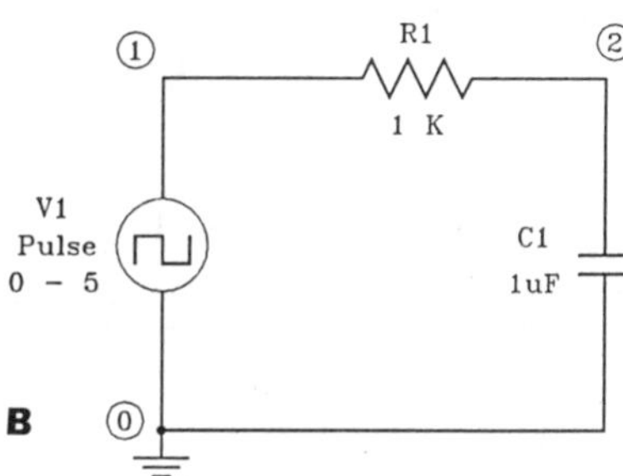

Example 2.5

the capacitor reaches full charge gives a time-dependent source. Dropping the source to 0 Volts simulates the shorting discharge path. The PULSE voltage source supplies the required square wave defined by the string of arguments within the parentheses. The general form for the PULSE source is:

PULSE (initial voltage
pulse voltage
delay time
rise time
fall time
pulse width
period)

Each of these arguments deserves a few words of explanation.

Initial Voltage = 0 This is the starting voltage and the low-level voltage of the square wave. In our example, the voltage applied to the circuit is 0 (zero) until the switch is moved to position A. At that time, the applied voltage jumps to 5 Volts.

Pulse Voltage = 5 This is the high-level voltage of the square wave. In our example, it is 5 Volts corresponding to the applied 5 Volt DC source.

Delay Time = 0 This is the amount of time between the beginning of our observation of the circuit and the beginning of the first pulse. In our example, we begin to observe the circuit at the instant when the switch is thrown. Consequently the delay is zero seconds.

Rise Time = 1N The rise time is the amount of time required for a signal to change from its low-level voltage to its high level. This must be a value greater than zero. Every action requires a finite amount of time, although the amount may be vanishingly small. Here the rise time is given a value of one nanosecond. While PSpice has trouble with an ideal square wave (0 second rise time), PSpice will allow an arbitrarily small rise time that can very closely approximate an ideal square wave.

Fall Time = 1N The fall time is the amount of time required for a signal to change from its high-level voltage to its low level. The fall time, like the rise time, must be greater than zero, but it can be arbitrarily small.

Pulse Width = 5M The pulse width is the length of time the signal will remain at the high voltage level. Five milliseconds was chosen in this example because the RC time constant for the circuit is one millisecond. This pulse width, in effect, leaves the switch in position A for exactly five time constants, just long enough to bring the capacitor to full charge.

Period = 10M The period is the time needed to complete one cycle of the square-wave signal. Since the pulse width is 5 milliseconds, a period of 10 milliseconds implies the signal is at the low-level voltage for the remaining 5 milliseconds. In this example, the discharge path of position B is equivalent to a zero voltage on the source, allowing the capacitor to discharge through V1. In either case the discharging time constant is the same. A 50% duty cycle was chosen because the charging and discharging time constants in this circuit are identical. PSpice does not require this and the analysis proceeds correctly for whatever duty cycle the user indicates by the pulse width and period values.

Line 4 describes the capacitor. Since the symbol "R" is used to indicate resistors, it is not surprising that PSpice uses "C" to indicate capacitors. As in the case of resistors, the next two numbers indicate the nodes for the position of the capacitor in the circuit. The value of the capacitor is one microfarad, which is written as 1U.

.TRAN tells PSpice that a time-based analysis is to be used. The general form for the .TRAN command is

.TRAN <sample time> <analysis duration>

PSpice calculates circuit values as often as PSpice determines it is necessary in order to make an accurate graphical display of the analysis. This frees the user from making the determination. While .PROBE makes good use of this information, there would be an overwhelming amount, of difficult to interpret data, if it were all printed in the .OUT file. The sample time of one millisecond in the example circuit indicates that only information calculated at one millisecond intervals should be printed to the output file. In short, the .Print TRAN statement makes use of the sample time parameter to restrict the printed data to manageable amounts, but .PROBE ignores the sample time and displays all of the calculated data. Even if the .Print TRAN statement is not included in the circuit description, and the sample time parameter is not used, the sample time value could not be omitted. Some value, less than the analysis duration, must be included to avoid a syntax error.

The .PRINT TRAN command tells PSpice to include values for the voltage across the capacitor V(2), the voltage across the resistor V(1,2), and the loop current I(R1), in the .OUT file. The data are given in the table at the top of page 27.

The PROBE program can be set to run automatically when the simulation is complete by setting the Auto-run box in the Analysis\Analysis Options window, or run manually by clicking the PROBE hot button. When examining a time-based signal, it is useful to begin by graphing the input waveform as a reference. Graphing V(2)

TIME	V(2)	V(1,2)	I(R1)
0.000E+00	0.000E+00	0.000E+00	0.000E+00
1.000E−03	3.156E+00	1.844E+00	1.844E−03
2.000E−03	4.324E+00	6.760E−01	6.760E−04
3.000E−03	4.752E+00	2.479E−01	2.479E−04
4.000E−03	4.909E+00	9.088E−02	9.088E−05
5.000E−03	4.967E+00	3.126E−02	3.126E−05
6.000E−03	1.830E+00	−1.830E+00	−1.830E−03
7.000E−03	6.710E−01	−6.710E−01	−6.710E−04
8.000E−03	2.460E−01	−2.460E−01	−2.460E−04
9.000E−03	9.020E−02	−9.020E−02	−9.020E−05
1.000E−02	3.300E−02	−3.100E−02	−3.100E−05

and V(1,2) against the source illustrates the reversal in polarity of the resistor voltage. To see the change in the circuit current over time, a second plot should be generated using "Plot__control" and adding the trace I(R1).

The final output generated by PROBE is shown in Figure 2.4.

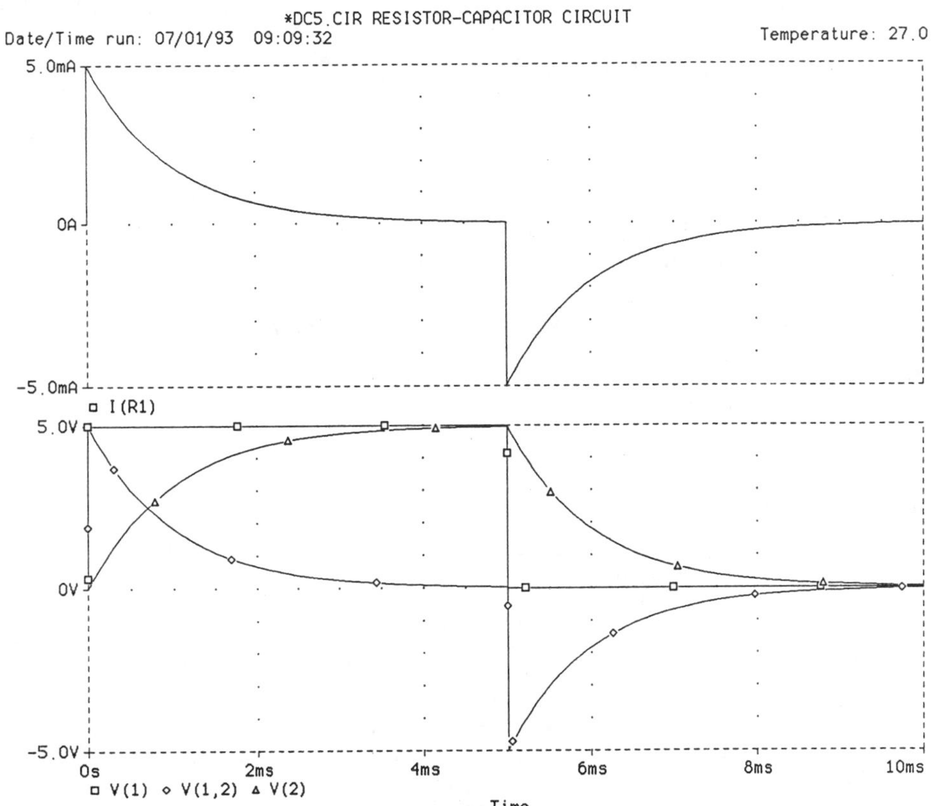

Figure 2.4

2.10 Transient Analysis of Resistor–Inductor Circuits

With a small modification, the same circuit can be used to test the transient response of a resistor–inductor circuit. By substitution of an inductor for the capacitor and introduction of a voltage-controlled switch, a circuit typically used as a lab exercise can be simulated. The symbol recognized by PSpice for an inductor is an "L," which is the standard symbol used in electronics.

Because an inductor's magnetic field must collapse as soon as the maintaining current is removed, circuits designed to measure an L/R time constant usually include a resistor such as R2 to provide a discharge path for the coil. Without this path, the current generated by the collapsing field will cause the switch to arc over. During the charging of the coil, R2 is in parallel with the source and does not affect the current through the coil or the voltages across R1 or the coil.

Modeling Example 2.6A requires a new element to effectively disconnect the source from the circuit. Merely reducing the voltage source to zero would indicate a short circuit across R2 and identical L/R time constants for the charging and discharging phases. To model Example 2.6A accurately, an open circuit must replace the source when the voltage drops to zero. The voltage-controlled switch in Example 2.6B will produce a virtual short when the source is at 5 Volts and an open when the source falls to zero.

Modeling the switch requires two lines in the circuit description.

S1 2 1 1 0 V_SWITCH

This tells PSpice that the switch (S1) is placed between nodes 2 and 1. The switch is controlled by the voltage between nodes 1 and 0 (i.e., the source). V_SWITCH is the arbitrarily chosen name for the switch. The general form for the switch is:

S<name> < + switch node> < − switch node> < + control node>
\+ ***< − control node> <switch name>***

.MODEL V_SWITCH VSWITCH

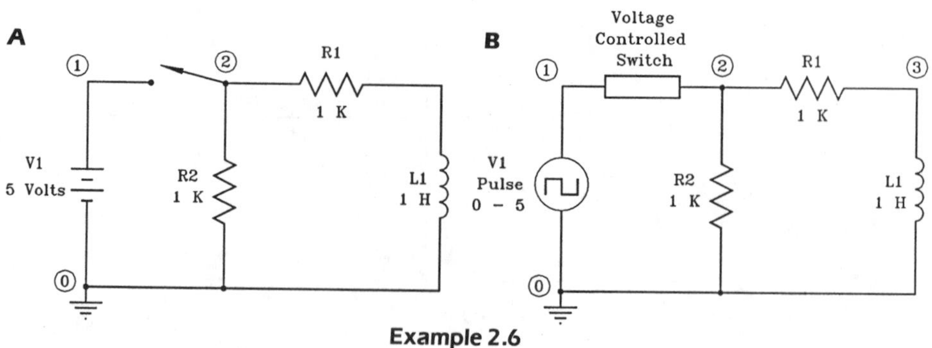

Example 2.6

Because PSpice can model many kinds of switches, it is necessary to specify the type of switch to use in the analysis. The .MODEL tells PSpice that the switch that has been named V_SWITCH is of type VSWITCH. The parameters of VSWITCH can be user specified, but in this example the default values are acceptable. The general form is

.MODEL <switch name> <switch type>

By default a VSWITCH has a value of 1 ohm when the controlling voltage is at least 1 Volt, and 1 Megohm when the controlling voltage is 0 volts or less.

The resulting circuit description is

```
*DC6.CIR   RESISTOR-INDUCTOR CIRCUIT
.MODEL V_SWITCH VSWITCH
V1 1 0 PULSE(0 5 0 1N 1N 5M 10M)
R1 2 3 1K
R2 2 0 1K
L1 3 0 1
S1 2 1 1 0 V_SWITCH
.TRAN .1M 10M
.PROBE
.END
```

Graphing the source voltage, the resistor voltage, the inductor voltage, and the circuit current yields textbook results.

2.11 Ideal Components

PSpice treats capacitors and inductors as ideal components. This means that capacitors have zero leakage current (infinite parallel resistance) and inductors have zero winding resistance. Since the parallel resistance of capacitors is usually very large (tens or hundreds of Megohms), omitting the parallel resistor in the modeled circuit creates an error between the model and reality that is insignificantly small. Omitting the winding resistance for a large inductor, on the other hand, can result in large errors. In the previous example, a one-Henry inductor might have a winding resistance of 200 ohms. Since this would represent 17% of the total circuit resistance, the charging and discharging times for the real circuit would not match the times generated by the PSpice model. The circuit description should be altered to include this resistance.

```
*DC6.CIR RESISTOR-INDUCTOR CIRCUIT
.MODEL V_SWITCH VSWITCH
V1 1 0 PULSE(0 5 0 1N 1N 5M 10M)
R1 2 3 1K
```

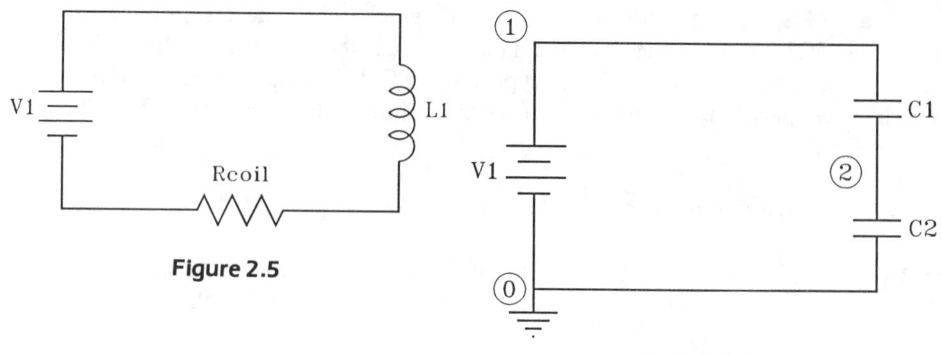

Figure 2.5

Figure 2.6

```
R2 2 0 1K
L1 3 4 1
RCOIL 4 0 200
S1 2 1 1 0 V_SWITCH
.TRAN .1M 10M
.PROBE
.END
```

Notice that the voltage across the inductor is still the voltage at node 3 and not V(3,4). In reality, it is impossible to take a voltage reading across the inductor without also measuring the voltage drop across the winding resistance.

Two other problems are caused by the ideal capacitor and inductor models used by PSpice. Consider the circuit shown in Figure 2.5.

Modeling this circuit without a value for the winding resistance results in an error when the analysis is attempted because PSpice interprets the inductor as zero resistance and a short across the source. For a less obvious reason, the capacitor circuit shown in Figure 2.6 could not be modeled without a parallel leakage resistance.

Again the problem is the treatment of the capacitors as ideal elements. In this circuit, node 2 represents an open in the circuit. PSpice must be able to establish a DC bias voltage for every point in the circuit. Since node 2 is isolated by two infinite resistances, no DC bias voltage can be established, and PSpice indicates an error. This problem can be avoided by including a parallel resistance for at least one of the capacitors. Adding a 100 Megohm resistor in parallel with C1 would not change the circuit performance appreciably, but would allow PSpice to establish the needed DC bias, and would make the circuit description more accurately model the real circuit.

2.12 Potentiometers

The only topic generally covered in a DC circuits analysis class is the use of variable resistors. PSpice can model potentiometers, but the model requires the use of subcircuits and other complex PSpice features that are inappropriate for inclusion at this point. Potentiometers are modeled in Chapter 5.

—3—

AC Circuits

The advantages of circuit analysis software increase dramatically when the voltage (or current) source changes in magnitude and polarity. Because capacitors and inductors cause phase shifts between the voltage and current in an AC circuit, the scalar algebra use to analyze DC circuits must be replaced by vector algebra. Not only do the calculations required to determine voltage levels and phase angles increase the difficulty of the analysis and the amount of time invested, but the repeated conversions between polar and rectangular coordinates greatly increase the opportunity for error.

PSpice not only eliminates the need for tedious calculation but can also display the results of the analysis in graphic form. It is generally true that numeric voltage and current levels in an AC circuit are less useful to a technician than the waveform those numbers describe. That is, technicians rely more on oscilloscopes when examining AC circuits than on volt meters. This is especially true at high frequencies.

In the previous chapter, PSpice is used to examine the voltage dropped across a capacitor charged by a DC supply. The table of time-dependent voltages was clearly less useful in understanding the behavior of the component than the same information in graphic form. Similarly, we rely on the ability of PSpice to graphically display circuit response in AC circuits.

For convenience, the syntax for descriptions of resistors, capacitors, and inductors is repeated here:

resistor

R<name> <pos node> <neg node> <value>

capacitor

C<name> <pos node> <neg node> <value>

inductor

L<name> <pos node> <neg node> <value>

PSpice allows the user to simulate many AC waveforms. The PULSE waveform is introduced in Chapter 2. Because a sine wave is the only waveform not distorted by capacitors and inductors, it is of primary interest in the study of AC circuits. Descriptions of other available waveforms, such as exponentials and piecewise linear waves, are found in Appendix C.

A sine wave voltage source is described by the following:

V<name> <pos node> <neg node> SIN (sine wave parameters)

First notice the use of positive and negative nodes. Although these terms appear to have little meaning for a source that is constantly changing its polarity, they in fact refer to the polarity at a reference point in time, which becomes more important when the circuit contains more than one source.

The sine wave parameters are described in the same fashion as the pulse wave parameters:

Sin ***(DC offset voltage***
peak amplitude
frequency
delay
damping factor
phase)

DC Offset Voltage This value is usually set to zero, but PSpice gives the user the opportunity to have the AC voltage riding on top of a DC biasing voltage.

Peak Amplitude PSpice interprets sine wave voltages as peak voltages rather than peak-to-peak or RMS values.

Frequency This value is the frequency of the sine wave measured in Hertz or cycles per second.

Delay PSpice allows the user to specify a delay in seconds between the time when examination of the circuit begins and the sine wave generator turns on. The delay is usually set to zero.

Damping Factor This value controls the exponential damping of the sine wave. The damping factor is generally set to zero but a precise value for damping can be calculated by using the formula given in Appendix C.

Phase The starting phase angle of the sine wave is usually set to zero as a matter of convenience, but this parameter is useful for solving problems that specify an initial phase angle and for multisource problems where the sources have a definite phase relationship. Phase angles are measured in degrees.

3.1 Series AC Circuit—Time-Based Analysis

Example 3.1 demonstrates the analysis of a series AC circuit containing a source, a resistor, and a capacitor. The component values have been chosen so that the capacitive reactance is equal to the resistance value, creating a 45° phase shift between the voltage and the current.

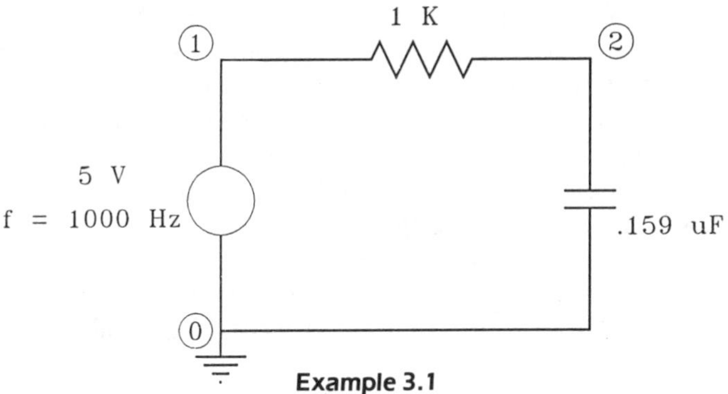

Example 3.1

```
*AC1.CIR—SERIES CIRCUIT      Time based analysis
V1   1 0 SIN(0 5 1000)
R1   1 2 1K
C1   2 0 .159U
.TRAN .1M 5M
.PRINT TRAN V(1) V(1,2) V(2)
.PROBE
.END
```

The circuit description begins, as always, with a title line that is ignored by the analysis program. The source is described as a sine wave that has zero DC offset, 5 Volt peak voltage, and a frequency of 1000 Hz. The remaining parameters are omitted; they are understood by PSpice to be zero. If a phase shift is required, values for the delay and damping factor must be included as place holders even if the values are zero.

```
V1   1 0 SIN(0 5 1000 0 0 90)
```

.TRAN indicates to PSpice that a time-based analysis is required. The first parameter of the .TRAN statement determines the print step value. In this case, PSpice has been asked to print a table of values in the .OUT file, with readings taken every .1 millisecond or every 36°. Even though we are more interested in seeing this information graphed and the .PRINT TRAN statement could be omitted, some value for the print step must be included in the .TRAN statement.

The second .TRAN parameter is more important. This value specifies how long the analysis should continue. Since the frequency of the AC waveform is 1000 Hz, the value 5M continues the analysis for 5 milliseconds or 5 cycles. This value is chosen somewhat arbitrarily by the user. A circuit containing capacitors or inductors will experience transient effects that may last for somewhere between 2 and 100 cycles, depending on the component values. Viewing the output for this circuit shows that the transient effects settle out quickly and that 5 cycles is sufficient.

.PROBE tells PSpice to display the results of the analysis graphically, converting the monitor into a simulated oscilloscope. The source voltage, the drops in voltage across the resistor and capacitor, and the current are displayed as shown in Figure 3.1.

The display confirms the lessons of AC circuit analysis concerning phase shifts and the need to sum voltage drops vectorially around a loop. It also, apparently, confirms the assumption that 5 cycles would be sufficient for the transient effects to settle out. If the user would like to examine one cycle of the output, the best choice is to alter the scale of the X-axis, asking PSpice to graph the output from 4 to 5 milliseconds. This is done by choosing the "X-axis" option from the menu at the bottom of the screen. When the new menu appears, choose "Set_range." At the "enter a

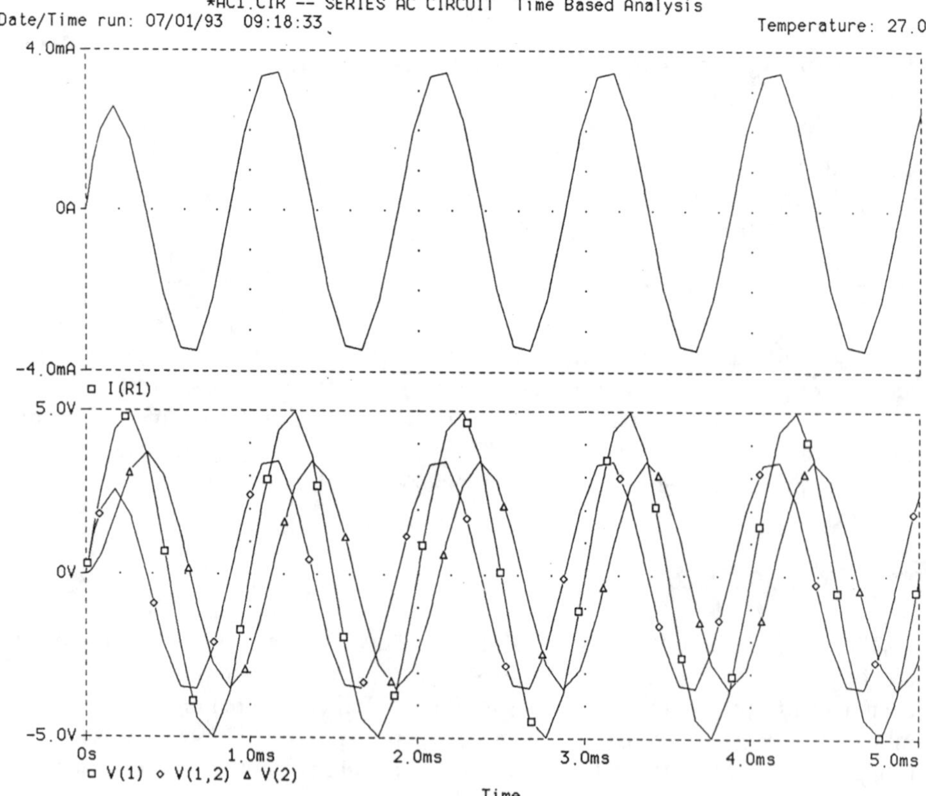

Figure 3.1

range" prompt type "4m 5m." This will cause the graph to be redrawn for the fifth cycle only.

> *CAUTION: In the PROBE program, the "m" indicating "milliseconds" is case sensitive. If "4M 5M" is used, PROBE will graph the output from 4 megaseconds to 5 megaseconds, yielding unexpected results.*

The rough contour of the graph may or may not be satisfactory. This results from the manner in which PSpice chooses the times at which the circuit is evaluated. PSpice plots as many points as it believes it needs to make a satisfactory plot. The default maximum time step is equal to the total analysis time divided by 50.

The .TRAN statement will accept two more optional parameters that can be used to improve the graphic output. The third value is the "no print" value. While the analysis begins at time zero seconds, the display disregards the data until the "no print" time is reached. In the .TRAN statement that follows, this has been set to 4 milliseconds to allow the circuit to "settle." That is, the first few cycles, which may display transient effects, are ignored. A smaller maximum time step can be set as the fourth argument. This tells PSpice the maximum interval that may exist between evaluated points in time. This results in more data points and a smoother contour to the plot. This value can be made arbitrarily small, but requesting a very large number of data points uses large amounts of computer memory and makes the analysis take longer.

```
.TRAN .1M 5M 4M .01M
```

Adding these extra parameters and rerunning the analysis software results in the graph shown in Figure 3.2.

The numeric values called for in the .PRINT TRAN statement are found in the .OUT file in tabled form. The following table is the result of the second analysis using the 4 millisecond "no print" value and the .01 millisecond step ceiling value. Notice the time interval between points is .1 millisecond (i.e., the first argument in the .TRAN statement).

TRANSIENT ANALYSIS TEMPERATURE = 27.000 DEG C

TIME	V(1)	V(1,2)	V(2)
4.000E−03	9.425E−10	2.500E+00	−2.500E+00
4.100E−03	2.937E+00	3.489E+00	−5.518E−01
4.200E−03	4.753E+00	3.147E+00	1.606E+00
4.300E−03	4.753E+00	1.603E+00	3.150E+00
4.400E−03	2.937E+00	−5.538E−01	3.491E+00
4.500E−03	3.169E−06	−2.499E+00	2.499E+00
4.600E−03	−2.937E+00	−3.489E+00	5.519E−01
4.700E−03	−4.753E+00	−3.147E+00	−1.606E+00

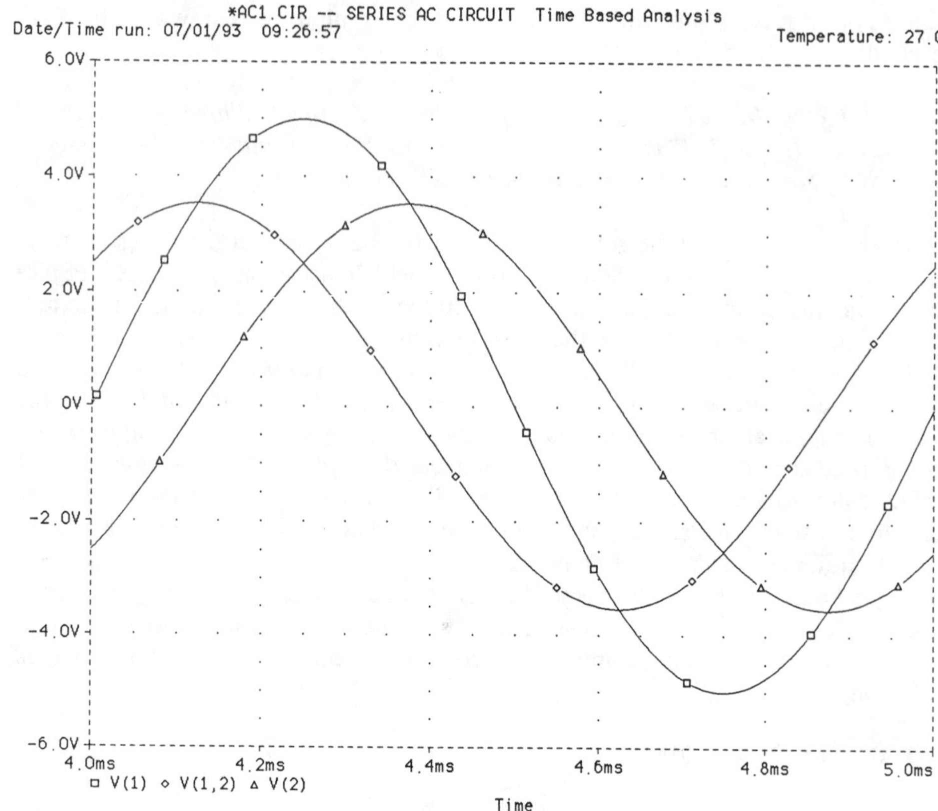

Figure 3.2

4.800E−03	−4.753E+00	−1.603E+00	−3.150E+00
4.900E−03	−2.937E+00	5.537E−01	−3.491E+00
5.000E−03	6.283E−10	2.500E+00	−2.500E+00

3.2 Series AC Circuit—Frequency-Based Analysis

PSpice can provide peak values for currents and voltage drops more conveniently by making a frequency-based analysis rather than a time-based analysis. The frequency-based analysis calculates a peak value without requiring the user to request a value at the exact time the waveform reaches its peak, which may be difficult to determine. In addition to a peak value, PSpice can provide phase angles relative to the source.

Frequency-based analysis of Example 3.1 requires a few alterations to the original circuit description:

```
*AC2.CIR—SERIES AC CIRCUIT   Frequency Based Analysis
V1 1 0 AC 5
R1 1 2 1K
C1 2 0 .159U
.AC LIN 1 1K 1K
.PRINT AC V(1) VM(1,2) VP(1,2) VM(2) VP(2) IM(R1) IP(R1)
.END
```

The first change is the description of the voltage source. The SIN( ) function is used only for time-based analysis. In this example the source is described as "AC 5." This indicates frequency-based analysis. A sine wave source is understood, and the value, 5, is taken to be the peak amplitude. The DC offset, delay, phase, and the like are all zero. The frequency is indicated in the .AC statement.

.AC signals PSpice to perform frequency analysis. LIN (linear), explained later in the discussion of filter circuits, pertains to frequency sweeping. "1 1K 1K" indicates the analysis should be carried out for one frequency, namely 1 kiloHertz. More specifically, one frequency value is to be used in the range beginning and ending at 1K. For filter analysis, we sweep many frequencies over a wide range.

The next line tells PSpice to print the results to the .OUT file. VM is the symbol for maximum voltage (i.e., Vpeak), and VP is the symbol for the voltage phase shift. The following data are taken from the .OUT file. Notice the voltage drops across the two components are not exactly the same, and the voltage phase shifts are not exactly 45°. This occurs because the capacitive reactance is not exactly 1K (1001.3 ohms).

FREQ	V(1)	VM(1,2)	VP(1,2)	VM(2)	VP(2)
1.000E+03	5.000E+00	3.534E+00	4.503E+01	3.537E+00	−4.497E+01

FREQ	IM(R1)	IP(R1)
1.000E+03	3.534E−03	4.503E+01

A few items that make the PSpice results intuitively satisfying are worth noting. As expected, the phase of the current through and the voltage across the resistor are identical. The current through and the voltage across the capacitor are 90° out of phase, with the current leading the voltage. Finally, the circuit current leads the source voltage in a capacitive circuit.

3.3 Multiple-Source–Multiple-Loop AC Circuits

Added circuit complexity greatly increases the amount of time and effort required to calculate component current and voltage drops by hand, but little extra effort is needed when using PSpice. Complex circuits involve extra care in naming nodes and writing the circuit description; otherwise, the procedures for simple and complex circuits are identical.

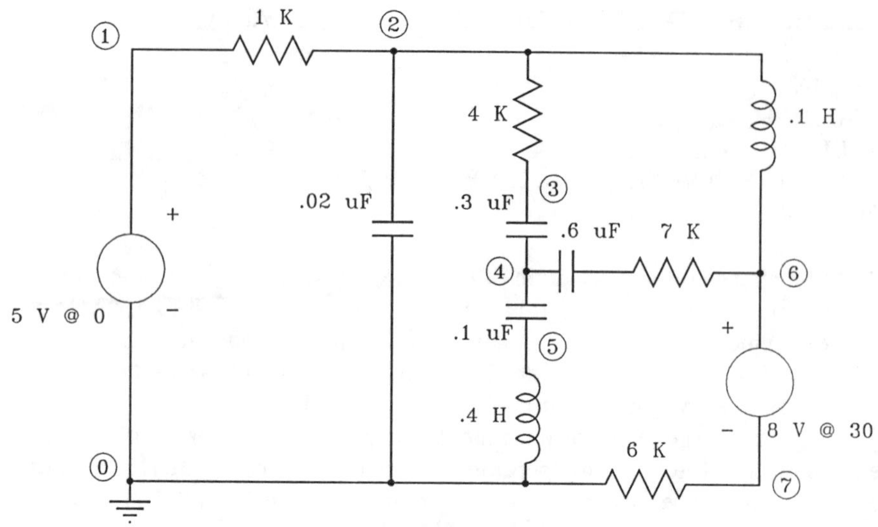

Example 3.2

Example 3.2 contains four loops and two sources that are 30° out of phase. While the numerically uninhibited may enjoy solving a system of four complex variable equations (or possibly using a delta-wye conversion and then Thevenizing), most would find determining the voltage drop across the 7K resistor an unhappy task. With PSpice the job is simple and quick.

In the following circuit description, the voltage sources are described first, then the resistor, capacitors, and inductors. The order for the listing of the elements is as arbitrary as their naming.

```
*AC3.CIR–MULTI-LOOP AC CIRCUIT
V1 1 0 SIN(0 5 1000)
V2 6 7 SIN(0 8 1000 0 0 30)
R1 1 2 1K
R2 2 3 4K
R3 5 6 7K
R4 7 0 6K
RCAP 4 5 100MEG
C1 2 0 .02U
C2 3 4 .3U
C3 4 5 .6U
C4 4 8 .1U
L1 2 6 .1
L2 8 0 .4
.TRAN .5M 5M 0 .01M
.PROBE
.END
```

Notice that in order to phase shift V2 by 30° it was necessary to include the zero delay and zero damping factor. Authors of circuit analysis textbooks are also fond of reversing the polarity of one of the voltage sources. To describe this to PSpice properly, be careful to list the positive node first on the line describing the voltage source.

The second item of interest is RCAP from node 4 to node 5. This resistor is not in the original circuit but must be included in the circuit description to allow PSpice to calculate a DC bias voltage for node 4. If this resistor is not included in the circuit description, PSpice will return an error message that node 4 is floating. This is a result of the ideal nature of capacitors in PSpice (i.e., capacitors have infinite dielectric resistance and zero leakage current). This feature isolates node 4 and stops the analysis.

The inclusion of RCAP as a 100 MEGOHM resistor is only slightly arbitrary. A real capacitor in the circuit will, in fact, have a finite resistance that is probably very large. The frequency-based analysis of this circuit that follows this discussion includes the values for the current through the capacitor and through the parallel resistance. Since the current through the capacitor is nearly 400,000 times larger than the current through the parallel resistance, it is clear that including RCAP does not significantly alter the circuit performance.

The .TRAN analysis would use .1 millisecond (5 millisecond / 50) as a default step ceiling and produce a poorly contoured graph. To smooth the graph, a step ceiling of .01 millisecond was included as the fourth parameter of the .TRAN line. The user should keep in mind that although this value can be made arbitrarily small, increasing the number of evaluation points increases the time required to analyze the circuit. In addition, beyond a certain number of evaluation points additional evaluations no longer improve the quality of the graph.

Figure 3.3 shows the two sources and the required waveform across the 7K resistor.

A new and useful feature has been added to the output graph. To obtain Figure 3.3, first plot the three waveforms. Then choose the "Cursor" menu option. The cursor option allows the user to choose a point on a particular waveform and obtain a numeric value for that point rather than estimating it. Maneuvering within the cursor menu using the keyboard is possible though rather awkward. A waveform and a position on the waveform can be indicated by interesting combinations of the <shift>, <ctrl>, and arrow keys, in addition to the capitalized letter in the menu options. This can more easily be learned by trial and error than by printed instructions. It is far more convenient to use a mouse to make choices. To obtain a peak value for the resistor voltage, click on the symbol to the left of V(5,6). Then click on the waveform near the fourth peak, click the "Peak" menu option, and finally click "Label_point."

For reasons about to be discussed, this is not the best way to obtain a peak value for the voltage across the 7K resistor. As in Example 3.1 a better choice is to use a frequency-based analysis. This involves altering the circuit description to read as follows:

```
*AC4.CIR—MULTI-LOOP AC CIRCUIT
V1 1 0 AC 5
```

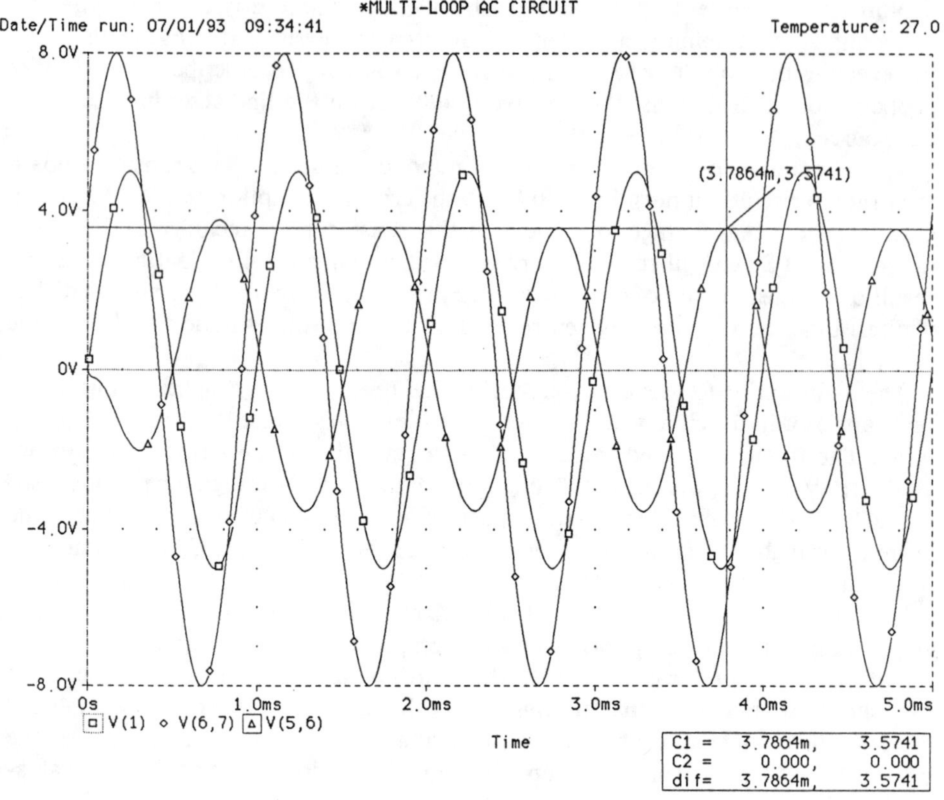

Figure 3.3

```
V2 6 7 AC 8 30
R1 1 2 1K
R2 2 3 4K
R3 5 6 7K
R4 7 0 6K
RCAP 4 5 100MEG
C1 2 0 .02U
C2 3 4 .3U
C3 4 5 .6U
C4 4 8 .1U
L1 2 6 .1
L2 8 0 .4
.AC LIN 1 1K 1K
.PRINT AC VM(5,6) VP(5,6) IM(RCAP) IM(C3)
.END
```

Compared to the frequency based analysis of Example 3.1, there is only one new item. The AC description for V2 has two parameters. The second is an optional parameter for a phase shift. If the parameter is omitted, as in V1, the value is assumed to be zero.

The .AC statement indicates that the circuit is be to analyzed at one frequency (1 kiloHertz). .PRINT AC requests the magnitude and phase for the voltage from node 5 to node 6. The last two values are the current values through the capacitor C3 and its parallel resistance.

The following are the values returned to the .OUT file:

```
****        AC ANALYSIS              TEMPERATURE =      27.000 DEG C
  FREQ        VM(5,6)       VP(5,6)       IM(RCAP)        IM(C3)
1.000E+03   3.560E+00    1.660E+02     1.349E-09     5.086E-04
```

There is a discrepancy between the value of 3.560 Volts given in the .OUT file and the 3.574 Volts given by the Probe software. 3.560 Volts is the correct value. The error in the peak value given by Probe results from the rash, but nearly true, assumption that all transient effects had been eliminated by the fourth cycle. Replacing the .TRAN parameters in the circuit description with

```
.TRAN .5M 20M 15M .01M
```

and requesting a peak value on the nineteenth cycle, gives better agreement with the AC analysis value.

3.4 Series Resonant Circuits

Series resonant circuits, in their most simple form, consist of a capacitor, an inductor, and some resistance. The resistive value of the circuit is the sum of the inductor's winding resistance and any other resistive components. Winding resistance is often ignored in circuit analysis at nonresonant frequencies because its value is usually small compared to the values of the other components. At resonance, however, the reactive values of the capacitor and coil sum to zero. For the resonant frequency, and for frequencies near resonance, the winding resistance may be a significant portion of the total circuit impedance.

In considering resonance, the focus of circuit analysis changes from investigating the voltage drop across each component to noticing the change in the amount of voltage each component drops as the frequency is varied over a certain range.

In an examination of Example 3.3, it is apparent from basic circuit theory that virtually all the source voltage is dropped across the capacitor at low frequencies. At high frequencies, nearly all the source voltage is dropped across the inductor. Since the capacitor voltage drop is 180° out of phase with the inductor voltage drop, there must be some intermediate frequency at which these drops cancel and all voltage drops across the resistive elements.

In the analysis of resonant circuits, we are usually interested in the circuit characteristics at and around the resonant frequency. The resonant frequency and the range of frequencies for which the circuit is primarily resistive (the bandwidth) can

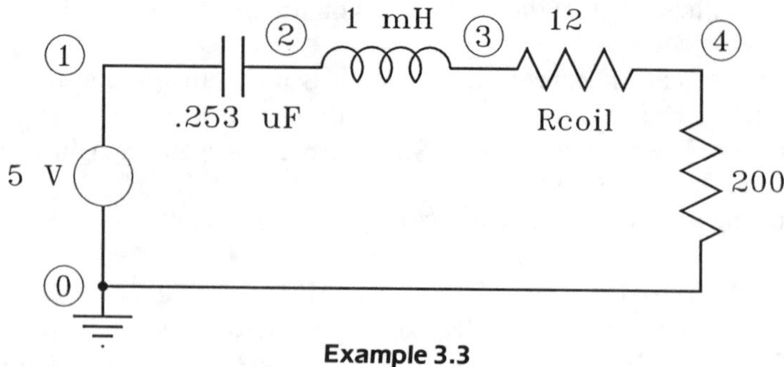

Example 3.3

be calculated mathematically. PSpice makes these calculations unnecessary by allowing the user to specify a large range of frequencies for the circuit analysis, and examining the graphic output. The resonant and cutoff frequencies can be determined by using PROBE to graph the circuit current or the load voltage.

The circuit description for Example 3.3 is

```
*AC5.CIR–SERIES RESONANT CIRCUIT
V1 1 0 AC 5
C1 1 2 .253U
L1 2 3 1M
RCOIL 3 4 12
RLOAD 4 0 200
.AC DEC 100 10 100MEG
.PROBE
.END
```

For PSpice to perform a frequency sweep, the source must be described as an AC voltage rather than a sine wave. In the .AC analysis statement, DEC (decade) is used as the first parameter to indicate that the output should be graphed on a logarithmic horizontal axis. The use of a logarithmic axis causes the frequency response for the load voltage to assume a bell-shaped curve. If LIN (linear) had been used as in previous examples, PROBE would have used a linear axis and the bell shape would have been distorted. The second parameter indicates that the circuit is to analyzed for 100 frequencies evenly spaced logarithmicly from 10 Hertz to 100 MegaHertz. LIN is more useful when making a table of values in the .OUT file.

Figure 3.4 is the graphic output of the Pspice analysis. For clarity, the "Plot_control" option was used to graph the output on three different plots. The topmost plot shows the load current. This indicates that the resonant frequency is 10 kHz. As expected, the center plot of the load voltage peaks at the same frequency. Notice that the load never drops the entire source voltage. The remaining few tenths of a volt drop across the winding resistance of the coil. The third plot shows the voltage drops for the capacitor and the coil (minus voltage dropped by the winding resistance). These values are, by definition, equal in magnitude at the resonant frequency.

Figure 3.4

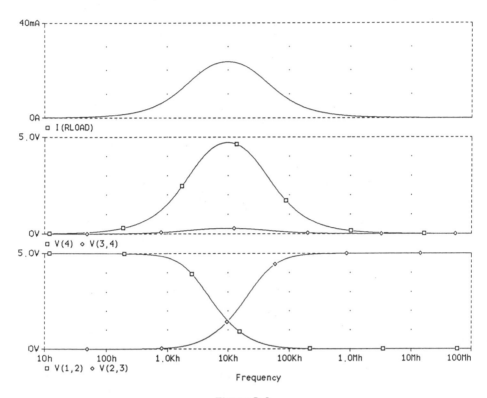

It is worth noting that in the laboratory, the values V(2,3) and V(3,4) are impossible to measure directly because a voltage measurement across the inductor necessarily includes both values.

To determine the cutoff frequencies and consequently the bandwidth, the following plot was constructed using the load voltage and altering the range of the axes. By using the "Cursor" option, locating the peak of the graph, and asking PSpice to label the point (as we did in the previous example), we learn that the resonant frequency is, in fact, 10 kiloHertz and that load voltage at that frequency is 4.7170 Volts. To determine the cutoff frequencies, first multiply 4.7170 by .707 to find the voltage at the cutoff frequencies. Then locate this voltage value on the graph and ask PSpice to label those points. By subtracting the provided cutoff frequencies, we obtain the bandwidth. The resulting graph is shown in Figure 3.5.

3.5 Parallel Resonant Circuits

Parallel resonant circuits are often found in the collector branch of transistor circuits. Since transistors are dependent current devices, most textbooks for AC circuit

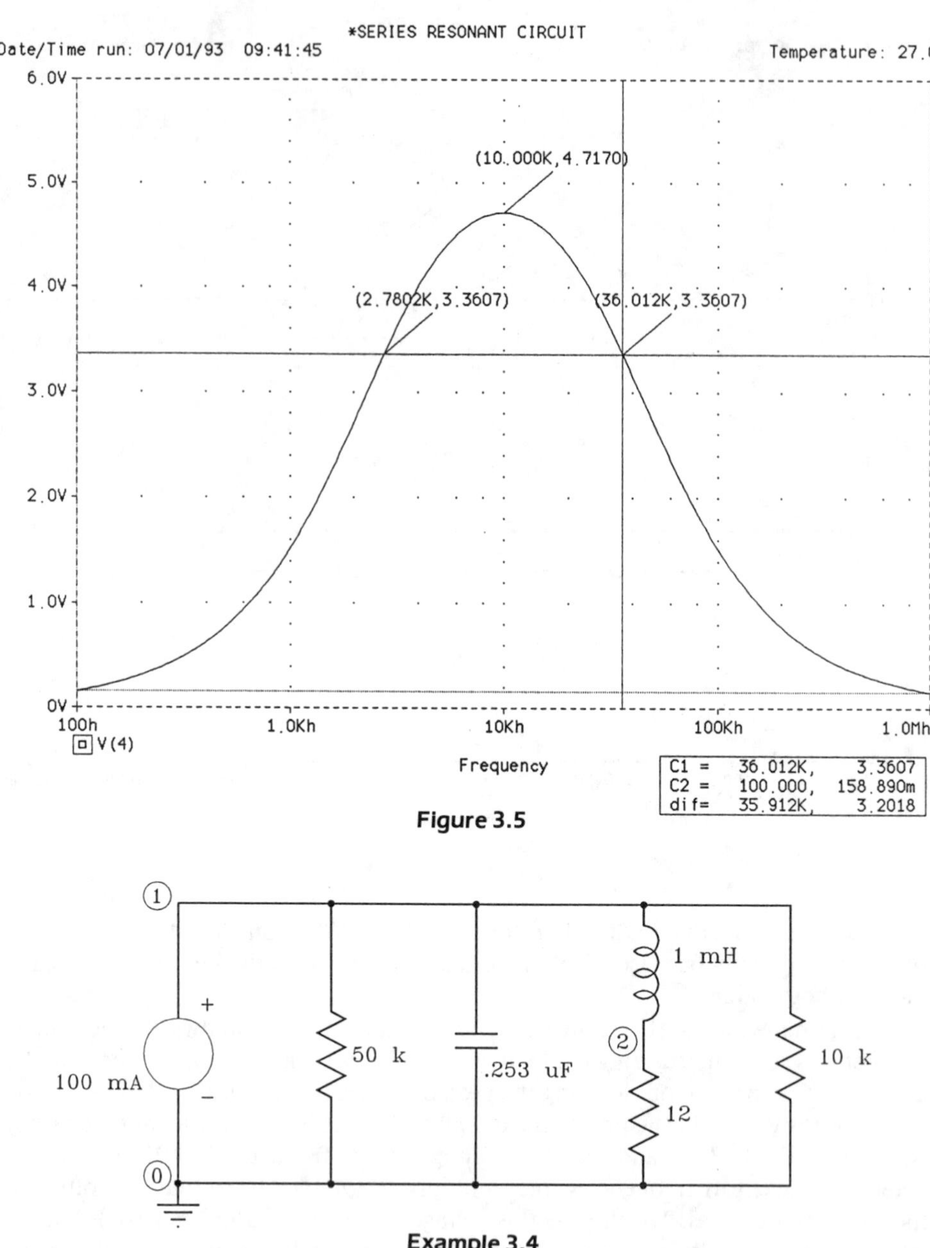

Figure 3.5

Example 3.4

analysis use a current source rather than a voltage source for problems involving parallel resonance. This is a reasonable convention, so Example 3.4 involves a current source, a tank circuit, and a 10K load resistor.

```
*AC6.CIR—PARALLEL RESONANT CIRCUIT
ISOURCE 1 0 AC 100M
```

```
RSOURCE 1 0 50K
L1      1 2 1M
RCOIL   2 0 12
C1      1 0 .253U
RLOAD   1 0 10K
.AC DEC 100 10 100MEG
.PROBE
.END
```

Except for the circuit configuration, there is little difference between Examples 3.3 and 3.4. A source resistance has been included for the current source because any value less than infinity broadens the bandwidth of the frequency response. The reactive component values and consequently the resonant frequency are the same for both examples.

Figure 3.6 shows the expected characteristics for a parallel resonant circuit. The topmost plot demonstrates the high current in the capacitor and coil loop. The center plot shows a much smaller current through the load. The third plot graphs the high voltages that can be generated across the current source. Each has the bell-

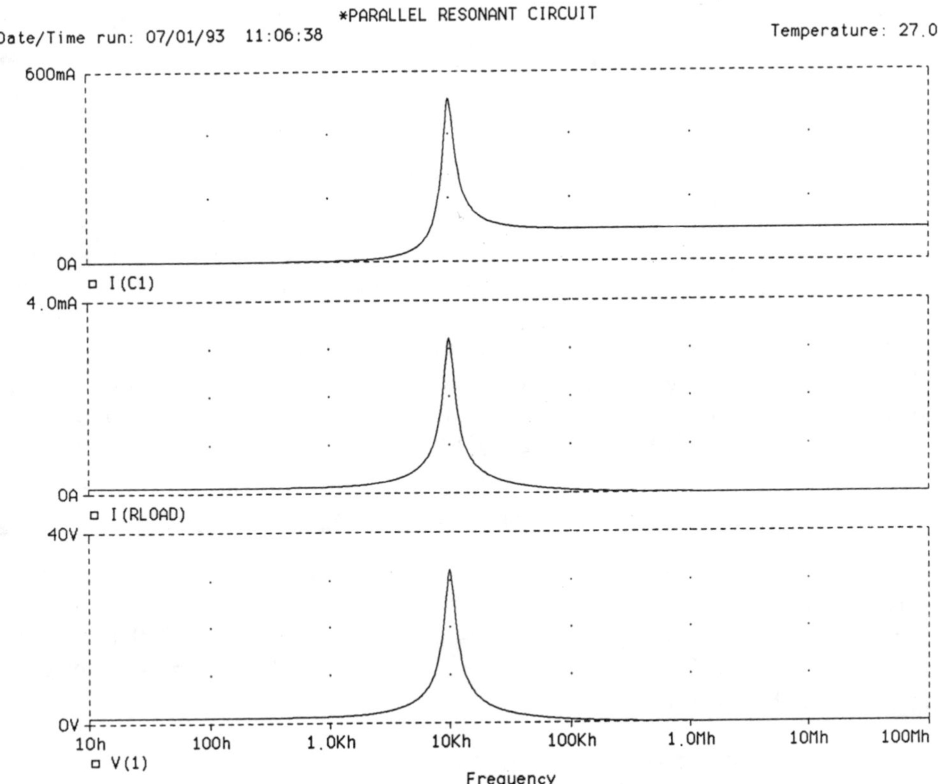

Figure 3.6

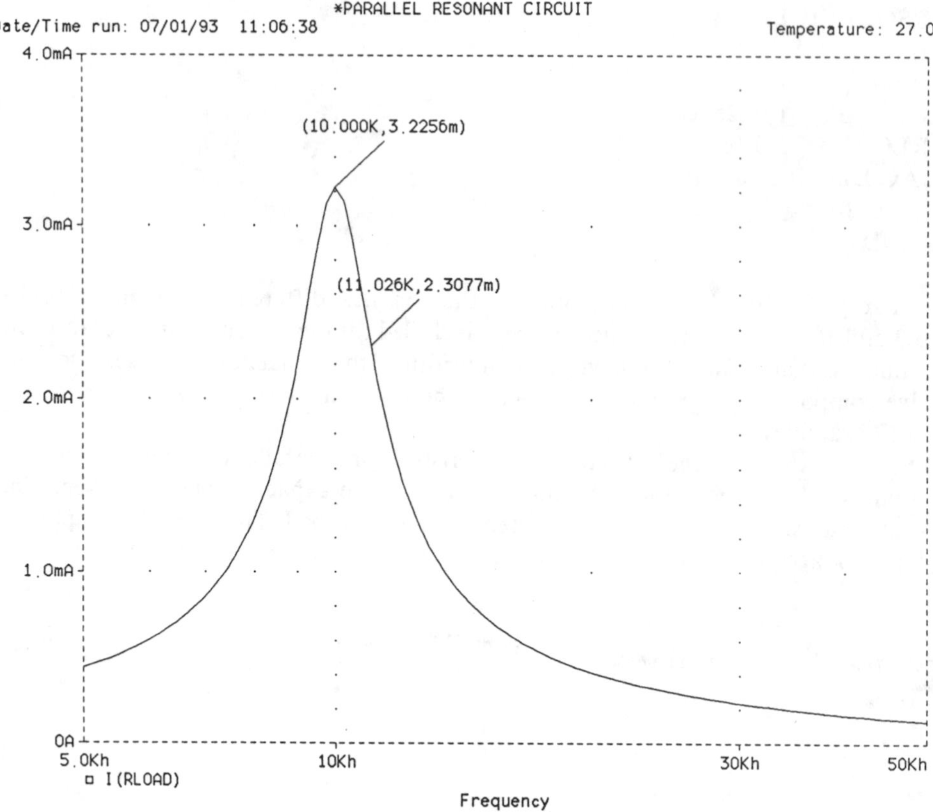

Figure 3.7

shaped frequency response seen in the series resonant circuit, centered at the same frequency.

Because parallel resonant circuits are usually designed for a narrow bandwidth, designers try to keep the load and the source resistances high. The bandwidth can be determined with the same procedure used for the series resonant circuit. Figure 3.7 indicates a bandwidth of approximately 2 kiloHertz.

3.6 High-Pass and Low-Pass Filter Circuits

The techniques developed for resonant circuits, which are generally used as band-pass filters, are equally applicable for high-pass and low-pass filters. Example 3.1, for example, can be configured as a low-pass filter, that is, a circuit that will pass low-frequency signals to the following stages but will short high frequencies to ground. See Figure 3.8.

```
*AC7.CIR—LOW PASS FILTER CIRCUIT
V1 1 0 AC 10
```

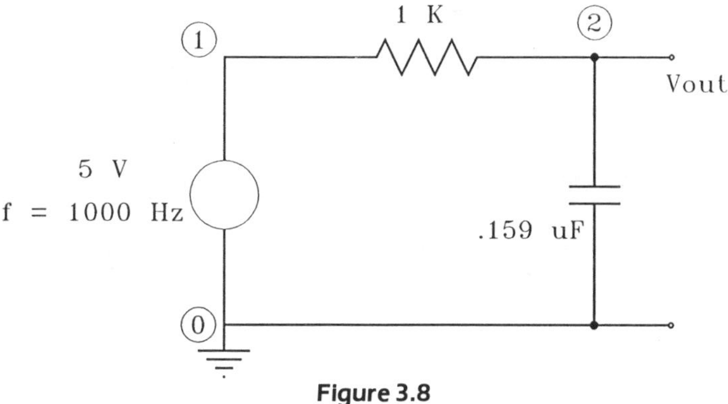

Figure 3.8

```
R1 1 2 1K
C1 2 0 .159U
.AC DEC 100 1 10MEG
.PROBE
.END
```

From a plot of the voltage generated across the capacitor, it is apparent that zero voltage is dropped for frequencies significantly higher than the cutoff frequency of 1 kiloHertz. Below the cutoff frequency, the reactance of the capacitor is greater than the resistor value. At frequencies lower than 100 Hertz, virtually all source voltage is dropped across the capacitor and passed on to the next stage. See Figure 3.9.

Example 3.5 is a multistage high-pass circuit. To pass high frequencies from stage to stage, the positions of the capacitor and the resistor have been interchanged compared to their positions in Figure 3.8.

```
*AC8.CIR—MULTI-STAGE HIGH PASS CIRCUIT
V1 1 0 AC 10
C1 1 2 .159U
R1 2 0 1K
C2 2 3 .159U
R2 3 0 1K
C3 3 4 .159U
R3 4 0 1K
.AC DEC 100 10 10MEG
.PROBE
.END
```

Filter theory (and intuition) indicate that with each added filtering stage the suppression of low frequencies should be greater. This implies that the slope of the frequency response curve, in the range near the cutoff frequency, should become steeper.

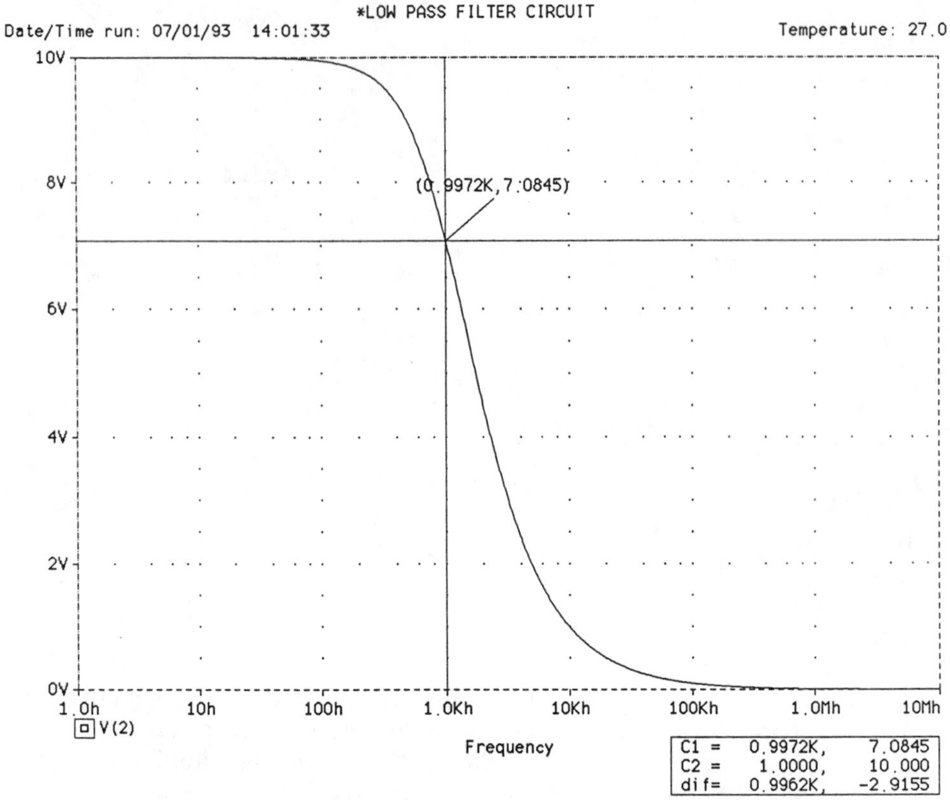

*LOW PASS FILTER CIRCUIT
Date/Time run: 07/01/93 14:01:33
Temperature: 27.0
10V
8V
6V
4V
2V
0V
(0.9972K,7.0845)
1.0h
10h
100h
1.0Kh
10Kh
100Kh
1.0Mh
10Mh
V(2)
Frequency
C1 = 0.9972K, 7.0845
C2 = 1.0000, 10.000
dif= 0.9962K, -2.9155

Figure 3.9

.159 uF
.159 uF
.159 uF
1
2
3
4
Vout
1 k
1 k
1 k
10 V
0

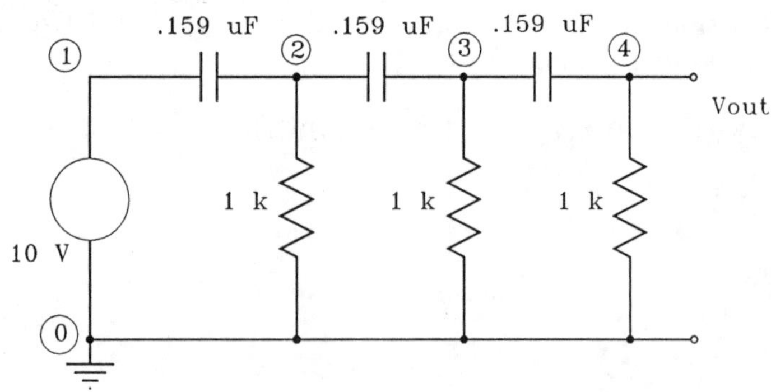

Example 3.5

By graphing the voltages at nodes 2, 3, and 4, PSpice can be used to verify the theoretical results. This is shown in Figure 3.10.

In practice, engineers usually speak of filters in terms of decibel suppression and graph frequency responses as Bode plots. A Bode plot is merely the frequency response curve graphed using a logarithmic scale for the Y-axis. The frequency response

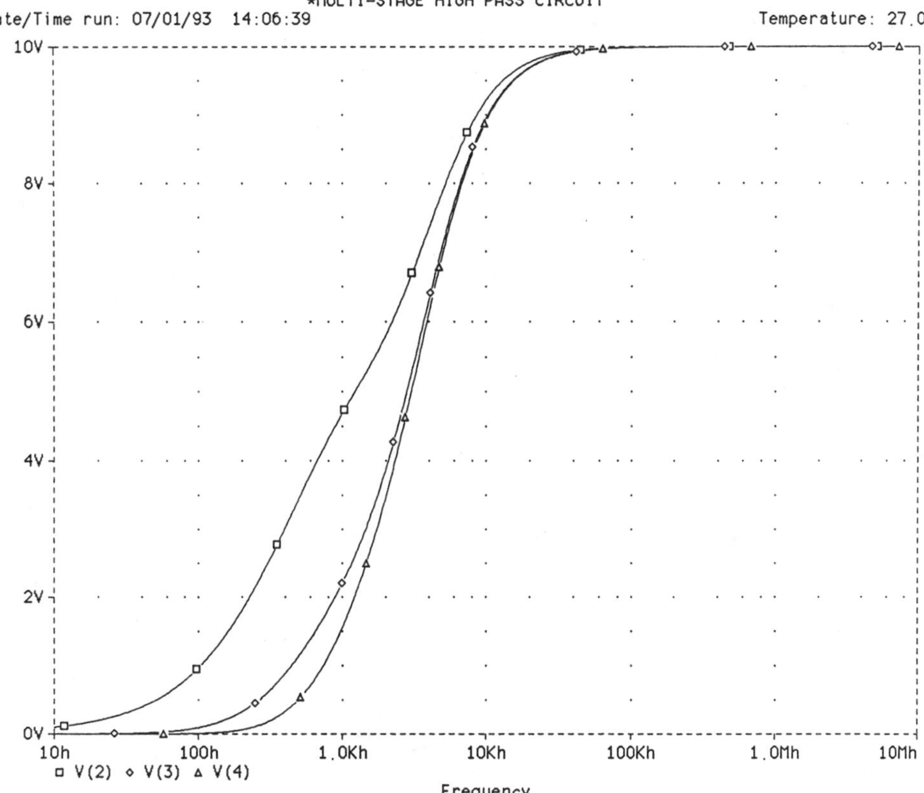

Figure 3.10

curve generated for Example 3.5 can be changed to a Bode plot by choosing the "Y-axis" option from the main PROBE menu and "Log" from the resulting submenu. The result of this operation is shown in Figure 3.11.

The straight line frequency response of the Bode plot below the cutoff frequency makes it much easier to read than the graph using a linear Y-axis.

By labeling strategic points and using the formula for decibel suppression, we can verify that at node 2 the first stage of the filter has suppressed the voltage by 20 dB. At node 4 the third stage has produced a 60 dB suppression. To avoid graphic clutter, the comparable points for node 3 were not labeled, but they would have shown a 40 dB suppression.

3.7 Transformer Circuits

With few exceptions, electrical power is generated and distributed as alternating current. Although direct current could be used, alternating current has one overwhelming advantage—it can easily be changed from one voltage level to another. Transformers can be used either to increase the voltage or to lower it. Power com-

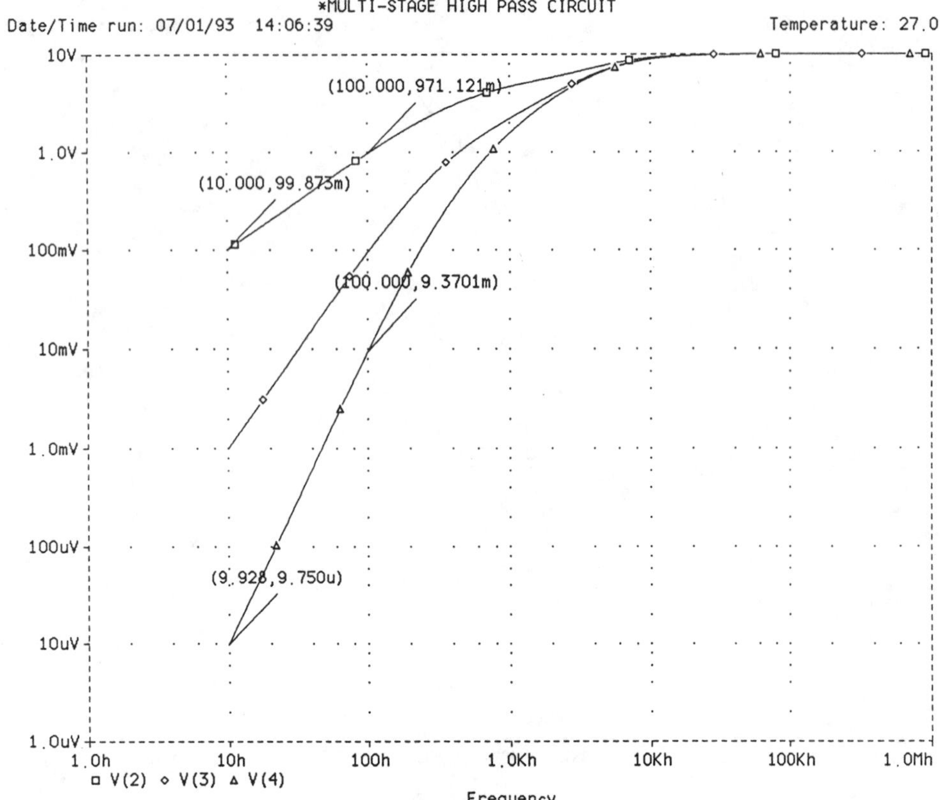

Figure 3.11

panies use step-up transformers to raise voltages and lower I^2R losses in long distance transmission. Step-down transformers reduce voltages to safer levels of 120 Volts and 240 Volts as the current enters our homes.

More widely found are the transformers that reduce the 120 Volts (often called line voltage) to even lower voltages. Step-down transformers are used for doorbell and low-voltage switching circuits and for the power supplies of our computers. This section demonstrates how PSpice models a 12.6 Volt step-down transformer. The nameplate data for small transformers are often limited to the frequency, which is generally 60 Hz, and the secondary voltage, assuming that the primary is connected to 120 Volts.

Since the values of 120 Volts on the primary coil and 12.6 Volts on the secondary are in RMS Volts, these values need to be converted to 170 Volts peak and 17.8 Volts peak, respectively, for PSpice modeling. In Example 3.6, the 1 K resistor is the load to be driven by the secondary voltage. The 42 ohm and 1 ohm resistor are typical for winding resistance values in a 12.6 Volt transformer.

In any transformer, the ratio of the primary voltage to the secondary voltage is the same as the ratio of the number of turns of wire on the primary to the number of

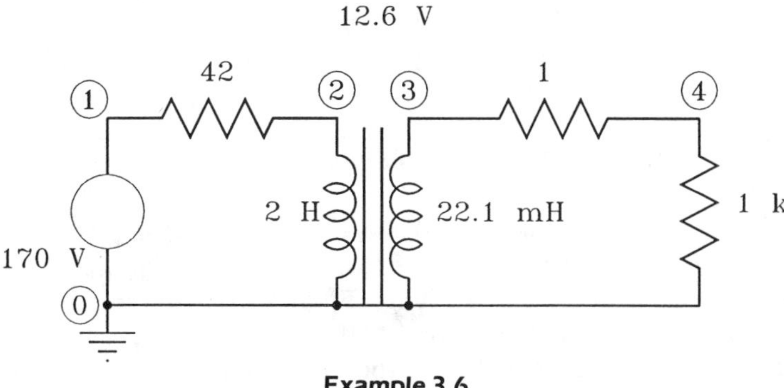

Example 3.6

turns on the secondary. This ratio is called the turns ratio. Because this is never specified on the nameplate it must be calculated.

alpha = 170 / 17.8 = 9.52

There is no single symbol for a transformer in PSpice as there are for resistors and capacitors. A transformer must be modeled from simpler components. One way to model a transformer is to use two tightly coupled coils. Unfortunately, there is no parameter in the description of coils for specifying the number of turns. To solve this problem, we start with the observation that the inductance of a coil is proportional to the number of turns squared. In this example, 9.524 is the required turns ratio, and 9.524 squared is 90.7, which is used as the ratio of the primary inductance to the secondary.

It is also important to notice that the secondary circuit has the same ground reference as the primary. PSpice will not accept a floating secondary for a transformer circuit modeled in this manner.

```
*AC9.CIR–TRANSFORMER CIRCUIT model 1
V1 1 0 AC 170
L1 2 0 2
R1 1 2 42
L2 3 0 .0221
R2 3 4 1
RLOAD 4 0 1K
K1 L1 L2 .999
.AC LIN 1 60 60
.PRINT AC V(1) V(1,2) V(4) I(V1) I(RLOAD)
.END
```

For the transformer in Example 3.6, the values for the coils are 2 Henry and .0221 Henry. These values were chosen because they are in a 90.7 : 1 ratio and because they are typical values for a 12.6 Volt transformer. The amount of coupling between the two coils is described by the line beginning with "K." The general form for the coefficient of coupling between coils is

K<name> <first coil> <second coil> <coupling value>

The coupling value must be between 0 and 1, with 0 indicating that no flux lines from the primary cut the secondary windings, and 1 indicating that all flux lines from the primary cut the secondary windings. For iron-core power transformers, the coupling coefficient is almost always greater than .99.

With one exception, the results of the PSpice analysis are the expected theoretical results.

*AC9.CIR—TRANSFORMER CIRCUIT model 1

****	AC ANALYSIS		TEMPERATURE =		27.000 DEG C
FREQ	V(1)	V(2)	V(4)	I(V1)	I(RLOAD)
6.000E+01	1.700E+02	1.697E+02	1.776E+01	2.250E−01	1.776E−02

Notice that V(1) and V(2) are nearly the same. This occurs because the 42 ohms value of winding resistance is small compared to 754 ohms for the reactance of the coil. The voltage and current values for the load are the expected 17.8 Volts and 17.8 milliamps. The erroneous value in this model is 225 milliamps for the primary current. This value comes from dividing 170 Volts by 754 ohms. The expected result is the secondary current divided by the turns ratio or 1.87 milliamps. This indicates that the load resistance was not reflected back through the transformer to the primary circuit.

To obtain a correct value for the current in the primary circuit, it is necessary to remodel the circuit. In Example 3.7, all elements of the secondary circuit have been multiplied by the turns ratio squared (90.7) and placed in the primary circuit.

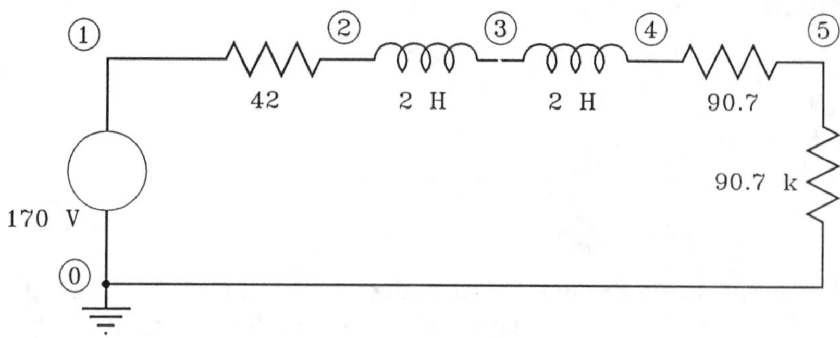

Example 3.7

The new circuit description and analysis are

```
*AC10.CIR—TRANSFORMER CIRCUIT model 2
V1 1 0 AC 170
L1 2 3 2
R1 1 2 42
L2 3 4 2
R2 4 5 90.7
RLOAD 5 0 90.7K
.AC LIN 1 60 60
.PRINT AC V(1) I(RLOAD) I(V1)
.END

****        AC ANALYSIS            TEMPERATURE =     27.000 DEG C

FREQ          V(1)            I(RLOAD)        I(V1)

6.000E+01     1.700E+02     1.871E-03     1.871E-03
```

This model gives the correct current supplied by the voltage source. The current for the load in this model is not the correct current for the original circuit load.

> *CAUTION: The value for the source current in this model is a theoretically correct, textbook value that does not correspond very well to reality. The true, experimentally determined, primary current value would vary from transformer to transformer, but it is likely to be in the range of 30 mA to 40 mA. This is attributable to several factors that have been omitted for the sake of simplicity. By including such elements as core resistance, magnetizing inductance, and other sources of loss, the model could be improved, and the simulation would more closely match reality. Because values for these sources of loss are not easily determined, a simplified model was used.*

Example 3.8 illustrates two more useful features in transformer modeling.

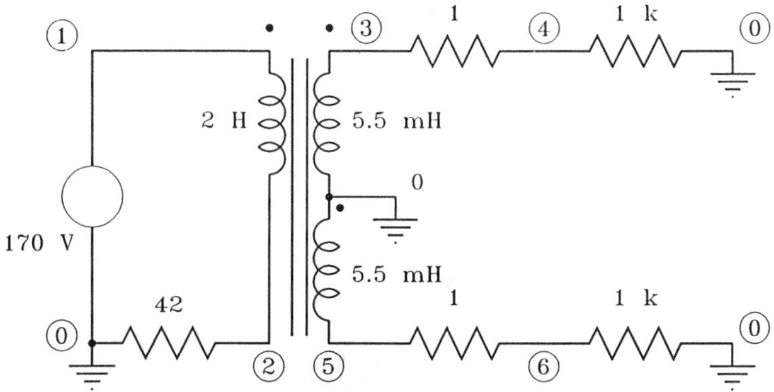

Example 3.8

```
*AC11.CIR–SPLIT SUPPLY
V1 0 1 SIN( 0 170 60)
L1 1 2 2
R1 2 0 42
L2 4 3 .0055
R2 3 0 1
RLOAD1 4 0 1K
L3 5 6 .0055
R3 5 0 1
RLOAD2 6 0 1K
K1 L1 L2 L3 .999
.TRAN .01 .08333 0 .1M
.PROBE
.END
```

This model pertains to a center-tapped transformer with two 6.3 Volt (RMS) secondary outputs that are 180 degrees out of phase. First notice that the size of the secondary coils is not one-half the inductance of those in Example 3.6 but one-quarter the inductance. A reduction by half of the secondary voltage implies a reduction by half of the number of secondary windings. Since inductance is proportional to the square of the number of windings, the inductance must be reduced by a factor of four.

In this example, the "K" line tightly couples three coils. In fact, any number of coils can be coupled by the "K" statement, making it possible to model transformers with many primary and secondary taps.

The last feature to note in the modeling of transformers is the use of the "dot" convention. It is common to indicate the phase relation for coupled coils by placing a dot on the schematic at the end of the coil where the current enters. PSpice uses this convention by interpreting the first node for the coil description as the location of the dot. The position of the dots (on the schematic and in the circuit description) indicates that the voltage of the source and RLOAD1 should be in phase and 180 degrees out phase with the voltage across RLOAD2. PROBE is used to verify the phase relationships and the proper amplitude for the load voltages. See Figure 3.12.

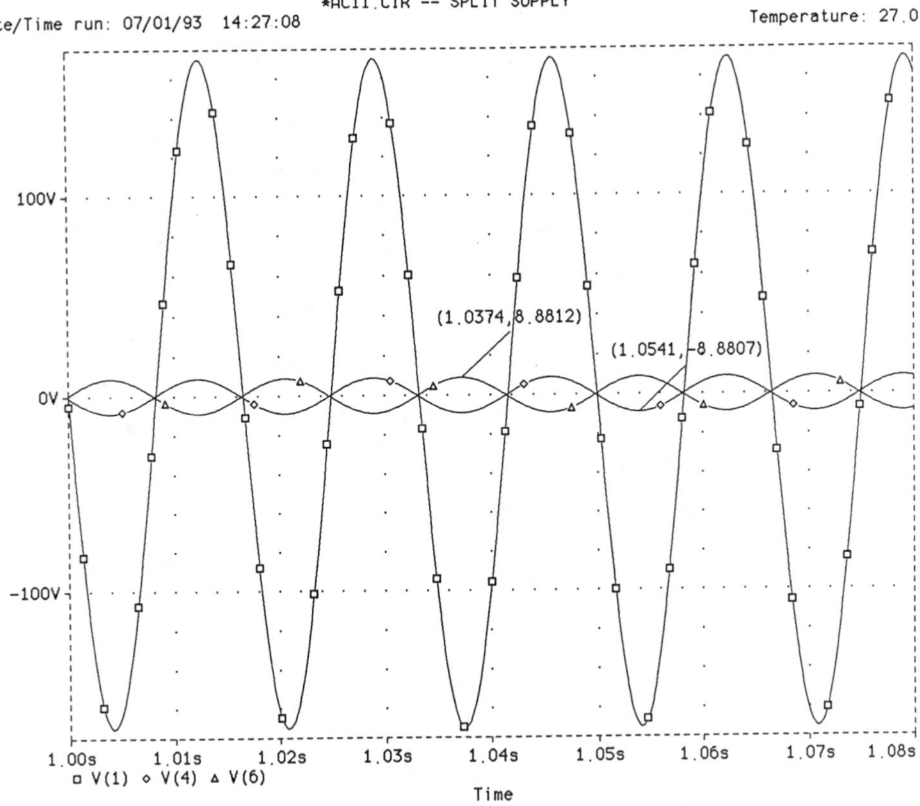
*AC11.CIR -- SPLIT SUPPLY
Date/Time run: 07/01/93 14:27:08
Temperature: 27.0
100V
0V
-100V
(1.0374, 8.8812)
(1.0541, -8.8807)
1.00s
1.01s
1.02s
1.03s
1.04s
1.05s
1.06s
1.07s
1.08s
□ V(1) ◇ V(4) △ V(6)
Time

Figure 3.12

4

Diode Circuits

The horizons of electronic circuitry broaden greatly with the introduction of semiconductor devices. Even the simplest semiconductor device, the single junction diode, makes possible rectification, regulation, and clipper circuits. In the correct configuration, diode circuits can even produce DC voltages at two, three, or more times the AC input voltage, making the need for large transformers unnecessary.

Because there are many different diodes with wide ranging characteristics, there is no generic diode that can be placed in a circuit description in the same fashion as resistors or inductors. There must be a way to indicate to PSpice exactly which diode is to be simulated. The properties of a 1N4006 are quite different from those of a 1N4148, and these switching diodes are markedly different from a 1N750 zener diode. Consequently, the writers of PSpice have created libraries of diodes (and transistors, op amps, and the like) to permit the user to specify exactly which diode is to be simulated.

In the evaluation version of PSpice, there is only one library. All diodes, op amps, and transistors have been modeled in EVAL.LIB. When a diode or other modeled component is used in a circuit description, information from the library containing the model must be available to the simulation programming. This is done in either of two ways.

The entire evaluation library can be made accessible by including the line .LIB \SPICE\EVAL.LIB in the circuit description. This line tells the PSpice software to search the evaluation library for any unrecognized device encountered in the circuit description. The needed information is placed in an index file (EVAL.IND). The index file is, in effect, a subset of the larger EVAL.LIB. This method has the advantage of convenience.

The second method involves the .INC statement. The .INC statement tells PSpice that the following file should be incorporated into the circuit description file. In previous versions of PSpice, attempts to include the entire EVAL.LIB (.INC\SPICE\EVAL.LIB) into a circuit description resulted in a "Symbol Table Overflow" error. Evaluation version 6.0, by using extended memory, allows the evaluation library to be included; however, including EVAL.LIB places the entire, rather large, file in the .OUT file of any circuit simulated. This requires the reader of the output file to scroll through many pages of unrelated information before finding the results of the simulation.

The solution is to create smaller libraries containing a few needed components. For this chapter, we create a library called DIODE.LIB containing only the diode models in EVAL.LIB. Although creating new libraries and including them with the .INC statement is slightly less convenient than using .LIB, this practice has two advantages.

Making a new library forces the user to know exactly which devices are in the library, that is, which models are available for use in a circuit description, rather than trying to use EVAL.LIB as a black box. The second advantage to using .INC is the inclusion of the library information in the .OUT file created by PSpice. Ready access to this information can be useful in making corrections to a circuit description that contains errors. The information defining diodes, transistors, and op amps is somewhat cryptic and more understandable to PSpice than to the user. Nevertheless, convenient subcircuits can be included in the library, and having subcircuit information in the .OUT file can be quite handy in trouble shooting a circuit description. The .LIB statement brings diode and transistor information into the .OUT file, but it does not insert subcircuit information.

In this chapter, DIODE.LIB is created and the .INC statement is used. This method was chosen because at the end of the chapter subcircuits are written for a 12.6 volt transformer and a bridge rectifier. Since these units are commonly used in circuits containing diodes, the subcircuit descriptions are placed in DIODE.LIB. When these subcircuits are used, their descriptions are placed in the .OUT file.

4.1 Creating DIODE.LIB

The EVAL.LIB should be kept unaltered as a reference library. To create a new library, change into the \SPICE subdirectory and use the DOS "copy" command to create a copy of EVAL.LIB called DIODE.LIB.

```
C:\SPICE>COPY EVAL.LIB DIODE.LIB
```

Then use an editor or a word processor to delete everything except the models for the diodes. The PSpice editor can be used, but there is a great deal to be deleted, and the PSpice editor scrolls rather slowly. To use the editor that comes with DOS 5.0 type

```
C:\SPICE>EDIT DIODE.LIB
```

Regardless of the editor used, delete everything except the following models, and then save the file.

```
*** Zener Diodes ***
.model D1N750     D(Is=880.5E-18 Rs=.25 Ikf=0 N=1 Xti=3 Eg=1.11
++    Cjo=175p M=.5516
+     Vj=.75 Fc=.5 Isr=1.859n Nr=2 Bv=4.7 Ibv=20.245m Nbv=1.6989
+     Ibvl=1.9556m Nbvl=14.976 Tbvl = -21.277u)
*     Motorola     pid=1N750   case=DO-35
```

```
*        89-9-18 gjg
*        Vz = 4.7 @ 20mA, Zz = 300 @ 1mA, Zz = 12.5 @ 5mA,
**       Zz =2.6   @ 20mA

*** Voltage-variable capacitance diodes
.model MV2201     D(Is=1.365p Rs=1 Ikf=0 N=1 Xti=3 Eg=1.11
++       Cjo=14.93p M=.4261
+        Vj=.75 Fc=.5 Isr=16.02p Nr=2 Bv=25 Ibv=10u)
*        Motorola     pid=MV2201   case=182-03
*        88-09-22 bam       creation

*** Switching Diodes ***
.model D1N4148     D(Is=0.1p Rs=16 CJO=2p Tt=12n Bv=100 Ibv=0.1p)
*        85-??-??      Original library

.model MBD101     D(Is=192.1p Rs=.1 Ikf=0 N=1 Xti=3 Eg=1.11
++       Cjo=893.8f M=98.29m
+        Vj=.75 Fc=.5 Isr=16.91n Nr=2 Bv=5 Ibv=10u)
*        Motorola     pid=MBD101   case=182-03
*        88-09-22 bam creation
```

4.2 Diode Characteristics

The information in the library models is in a cryptic form that can be deciphered with the abbreviation codes found in Appendix C. Values for "forward-bias depletion capacitance coefficient" and "IS temperature coefficient," however, are of more use to PSpice than to the student.

For the student, the most relevant information about the performance of the diode is in the current-voltage response curve. This curve illustrates the junction voltage and the bulk resistance for the device. The current-voltage curve can be developed by sweeping a DC voltage across the forward-biased diode. For the 1N4148 diode, the PSpice circuit description is

```
*DIODE1.CIR-1N4148 DIODE TEST
.INC \SPICE\DIODE.LIB
V1 1 0 5
D1 1 0 D1N4148
.DC V1 0 1 .01
.PROBE
.END
```

There are two new items to note in this circuit description. First, the .INC statement requires the use of the **full library name including the path and the extension.** Second, the general form for a diode is

D<name> <anode> <cathode> <model>

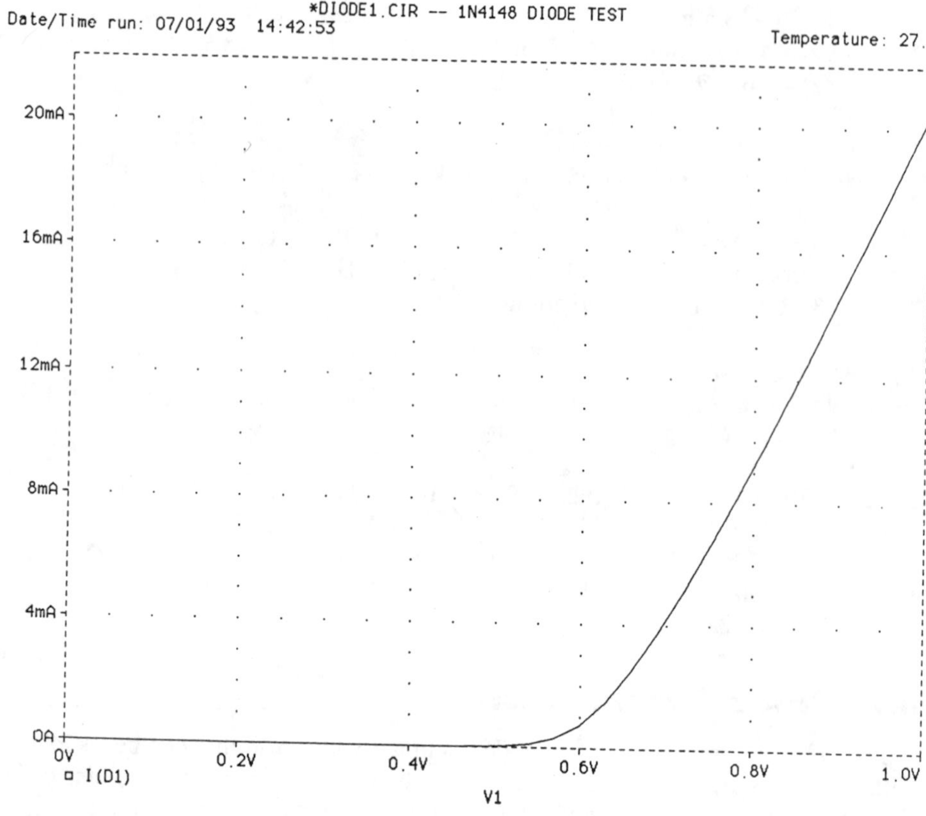

Figure 4.1

If the cathode node is listed first, PSpice will correctly analyze the circuit for a reversed biased diode. The DC sweep statement describes V1 increasing in voltage from zero to one volt in increments of .01 Volts. The sweep range overrides the 5 Volts indicated in the V1 description. PROBE is used for a graphic display of the results, as shown in Figure 4.1. Making a change in the diode model and the sweep range permits development of a similar response curve for the MBD101 diode. This is shown in Figure 4.2.

4.3 Half-Wave Rectifiers

In most technical semiconductor textbooks, the first AC diode circuit encountered by students is the half-wave rectifier. Since no new features are introduced in this example, the circuit (Example 4.1), the circuit description, and the graphic output (Figure 4.3) are presented without further comment.

```
*DIODE4.CIR—HALF WAVE RECTIFIER
.INC \SPICE\DIODE.LIB
```

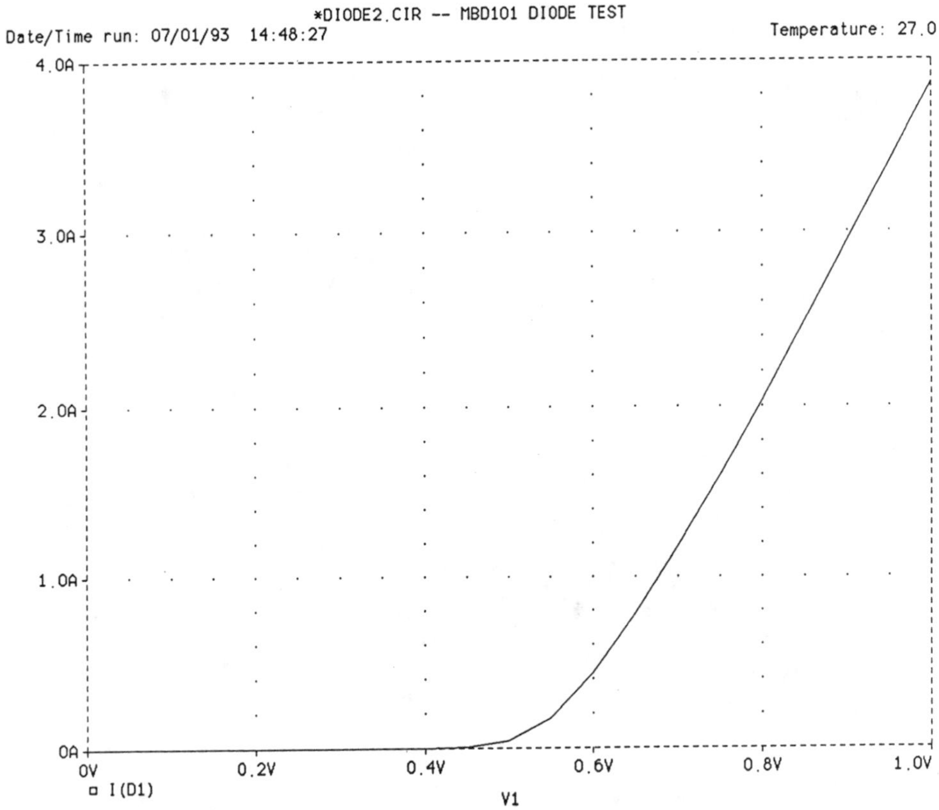

*DIODE2.CIR -- MBD101 DIODE TEST

Date/Time run: 07/01/93 14:48:27

Temperature: 27.0

4.0A

3.0A

2.0A

1.0A

0A

0V 0.2V 0.4V 0.6V 0.8V 1.0V

□ I(D1)

V1

Figure 4.2

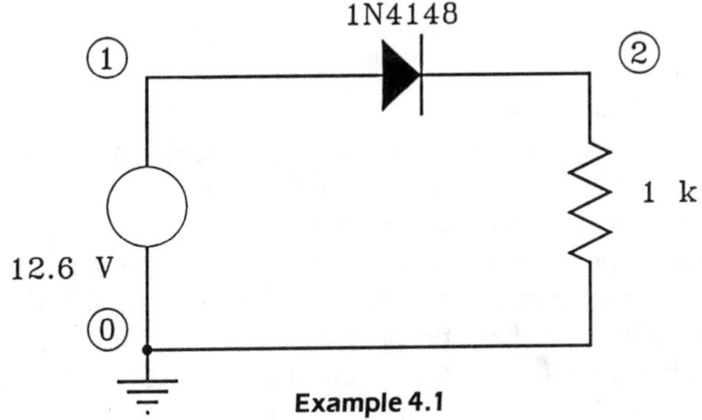

Example 4.1

```
V1 1 0 SIN ( 0 12.6 1000 )
D1 1 2 D1N4148
R1 2 0 1K
.TRAN .1M 10M 5M .01M
.PROBE
.END
```

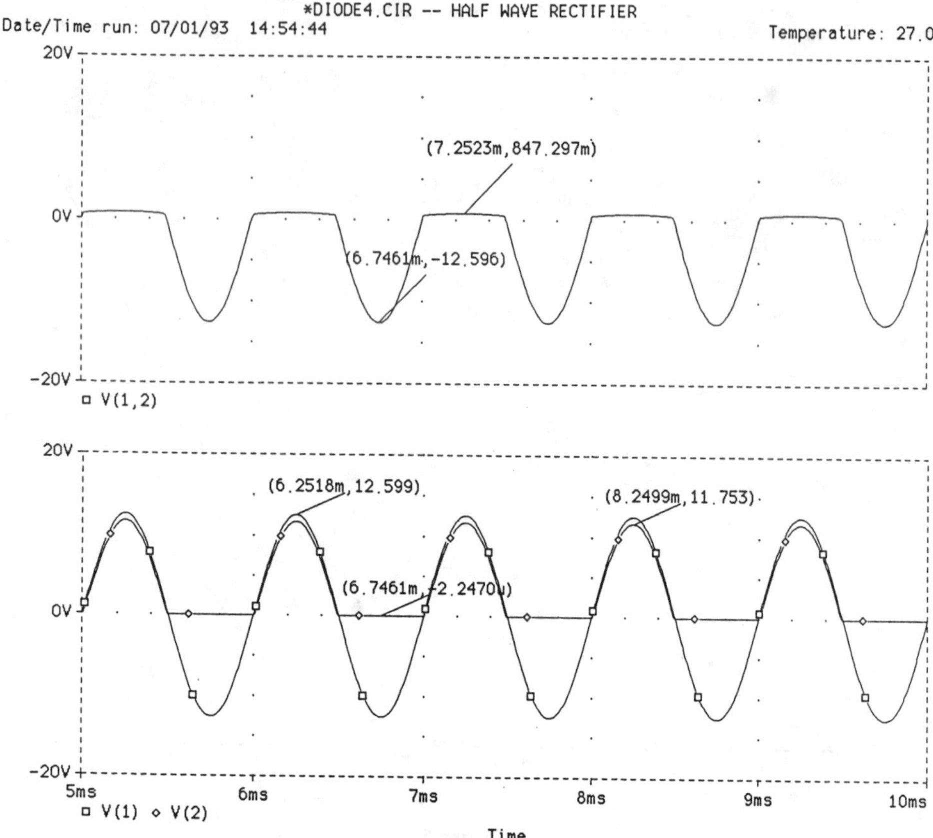

Figure 4.3

The upper plot in Figure 4.3 graphs the voltage dropped by the diode, and the lower plot graphs the voltage across the load and the source voltage as a reference. Through examination of the labeled points, several expected results become apparent. When the source negatively biases the diode, all source voltage is dropped by the diode. While this is not strictly true, the leakage current through the reverse biased diode is so small that the voltage dropped by the load is virtually zero.

When the diode is forward biased, the difference between the source voltage and the load voltage is .85 Volts. This is the voltage the diode is expected to drop, based on the above 1N4148 voltage-current diode curve. To verify this, notice that when the source is at its peak, 11.75 Volts drop across the 1 K resistor. This implies that 11.75 mA of current is flowing through the resistor and through the diode. Tracing the diode curve to 11.75 mA on the Y-axis gives .85 volts on the X-axis.

4.4 Bridge Rectifiers

A bridge rectifier is a full-wave rectifier. In Example 4.2, diodes in a bridge configuration are used to change the polarity of the negative half of the input waveform rather than blocking the negative half, as occurred in the half-wave rectifier. The result is a pulsing DC wave at twice the frequency of the original input and twice the frequency that would result from using a half-wave rectifier.

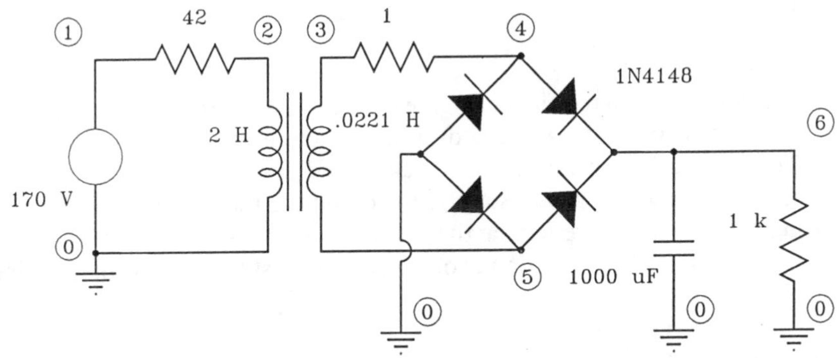

Example 4.2

Every piece of digital equipment requires a power supply to convert the 120 Volts (RMS) provided by the power company to a lower DC voltage to power IC chips. Because the vast majority of power supplies use a bridge rectifier, the schematic in Example 4.2 (although not yet complete) is very much a real world circuit. And since, in the real world we plug our computers into the wall outlet, a 170 Volt supply and step down transformer have been included in the circuit description.

In order to examine the unfiltered and the filtered output, an asterisk is included at the beginning of the line describing the capacitor (C1). In the following circuit description, C1 is a filter capacitor and the asterisk makes the capacitor description into a comment line. PSpice will analyze the circuit twice. The first analysis is with the asterisk (i.e., the unfiltered circuit with the capacitor removed). The second time, the asterisk is removed and the filter cap included.

```
*DIODE5.CIR—FULL WAVE RECTIFIER WITH FILTER
.INC \SPICE\DIODE.LIB
V1 1 0 SIN( 0 170 60 )
RPRIMARY 1 2 42
LPRIMARY 2 0 2
LSECONDARY 3 5 .0221
```

```
K1 LPRIMARY LSECONDARY .999
RSECONDARY 3 4 1
D1 4 6 D1N4148
D2 0 4 D1N4148
D3 0 5 D1N4148
D4 5 6 D1N4148
*C1 6 0 1000U
RLOAD 6 0 1K
.TRAN .016 .32 0 .60M
.PROBE
.END
```

The output for the unfiltered circuit (Figure 4.4) shows 17.8 Volts peak at the secondary (12.6 Volts RMS) and a voltage drop across the load that is two diode drops lower. The pulsing DC current from the unfiltered circuit has only a few applications. In a more practical power supply, the current is filtered and later regulated. Removing the asterisk and rerunning the PSpice analysis results in the output voltage shown in Figure 4.5. Clearly the load voltage, although still rising, is approaching an

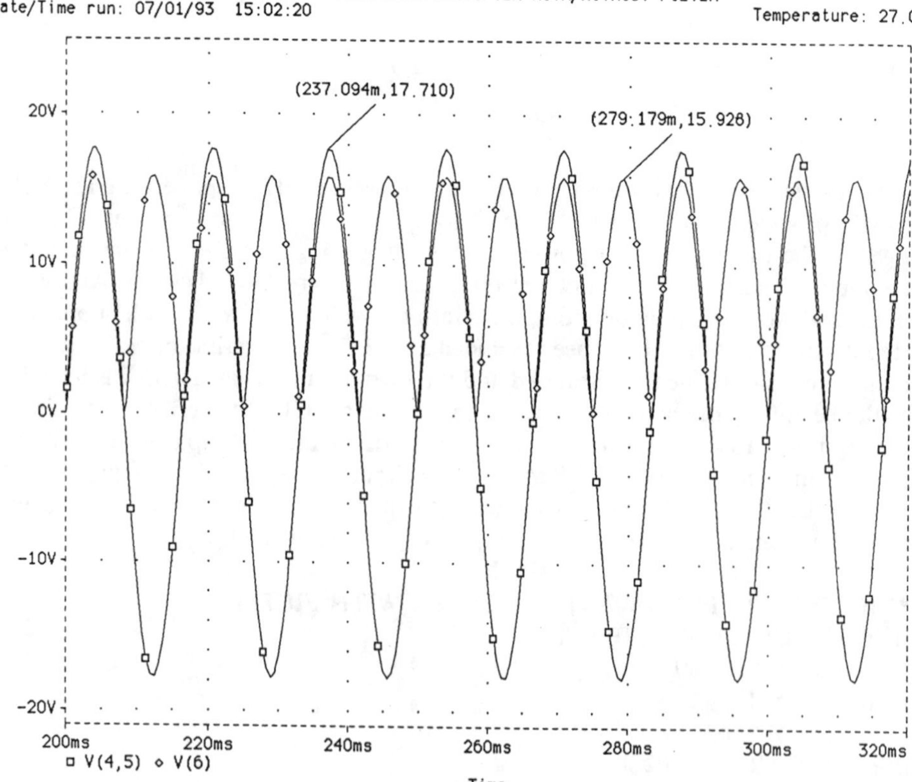

Figure 4.4

asymptote at around 14 Volts. By an adjustment in the ranges of the axes, or, more conveniently, use of the "Zoom" menu option and creation of a window with the mouse, Figure 4.6 can be developed. The results of the filtered circuit analysis may come as a surprise. Most textbooks indicate that the capacitor charges to a value equal to the peak load voltage of the unfiltered circuit, in this case 16 Volts. PSpice indicates that the load voltage is 2 Volts lower. The 16 Volt textbook value is based on the assumption that a forward-biased diode drops about .7 Volts. Sometimes this assumption is reasonable and sometimes it can lead to considerable error. PSpice uses a much more accurate model for its diodes. It is clear from the diode curve for the 1N4148 that the amount of voltage drop depends on the amount of current flowing through the diode. The plot in Figure 4.7 graphs the current flowing through the capacitor and the load resistor.

Adding the filter capacitor increases the current through the diodes from 14 mA for the resistor only to almost 75 mA for the peak current through the load and the capacitor. If we were to have PSpice recreate the 1N4148 diode curve for the range 0 to 3 Volts, we would find that at 75 mA each diode drops 1.9 Volts. Across two diodes the drop is 3.8 Volts, and the difference between the transformer secondary voltage and the load voltage is explained.

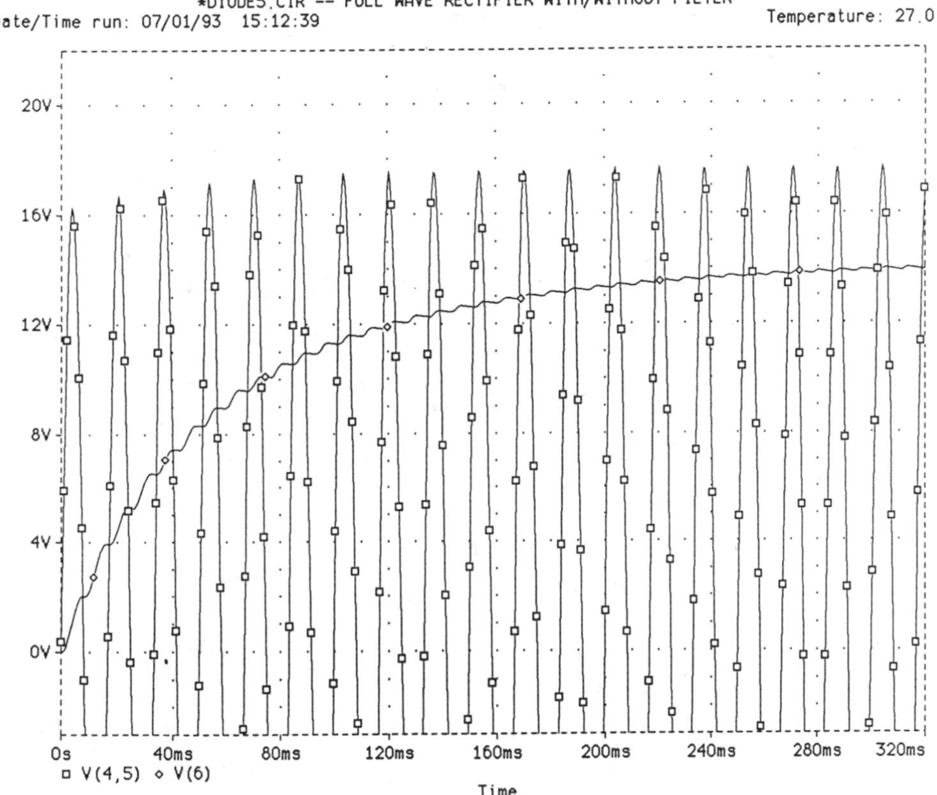

Figure 4.5

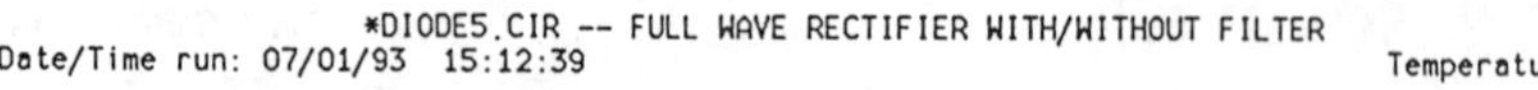

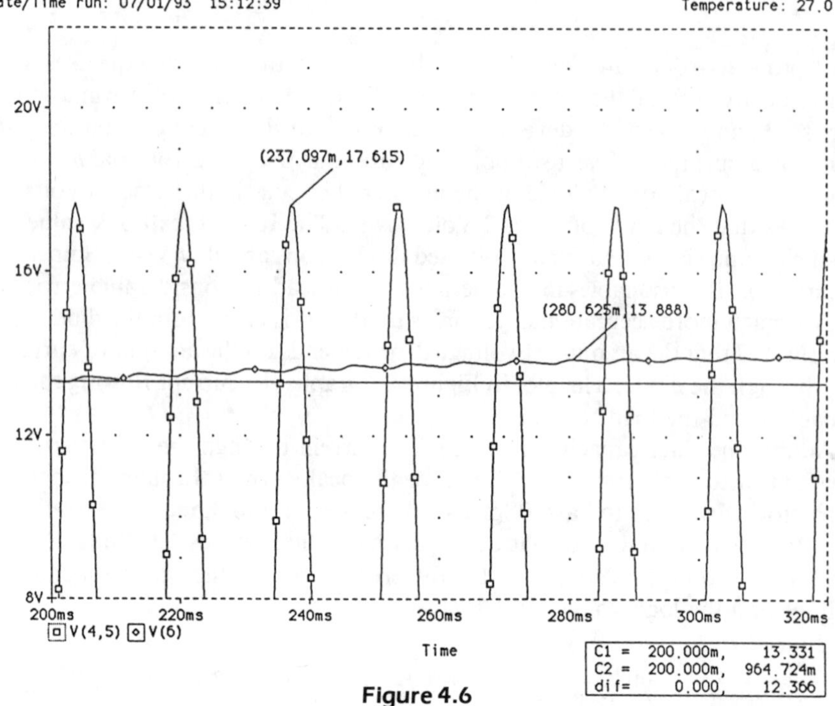

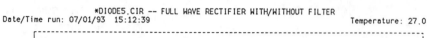

Figure 4.6

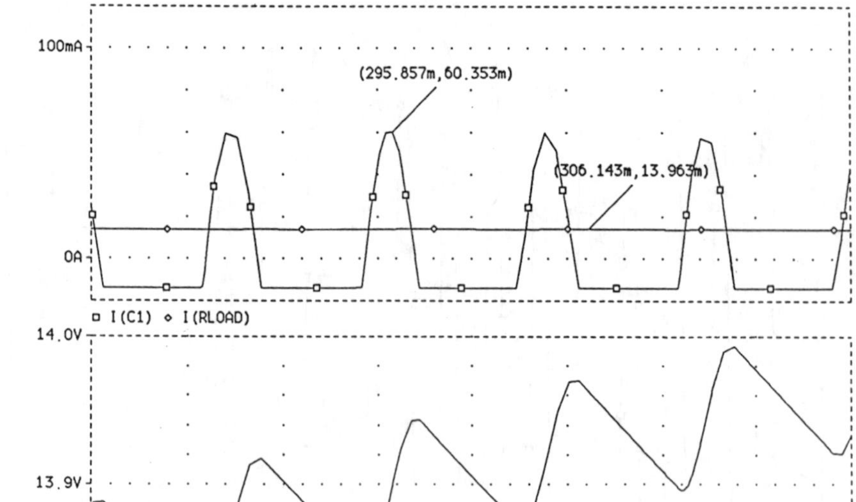

Figure 4.7

4.5 Power Supplies

Once the transformer output is rectified and filtered, the next reasonable step is to regulate the load voltage. In practice, power supplies use integrated circuit regulators, but nearly the same result can be obtained using a zener regulator diode. This is fortunate because no IC regulators are included in the evaluation library. IC regulators have largely replaced zeners for voltage regulation because of their built-in overcurrent protection and improved ripple voltage reduction. Zeners are still used as preregulators to improve the performance of following IC regulators.

By inserting a 1N750 zener diode, the previous circuit can be altered to reduce the 14 Volt output of the filtered rectifier to 4.7 Volts with considerably less ripple, as shown in Example 4.3.

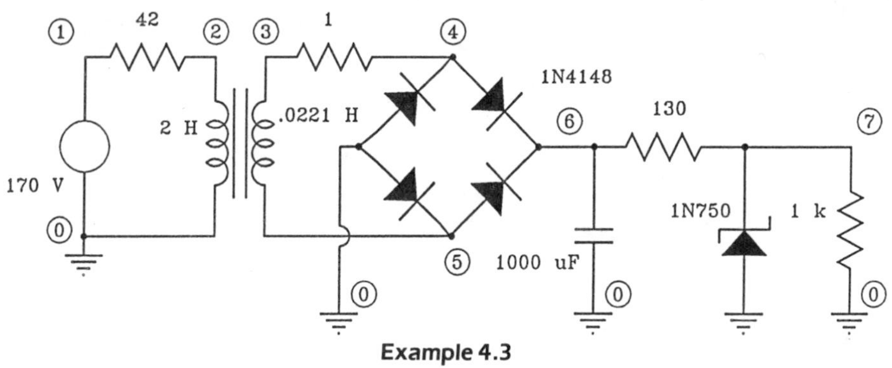

Example 4.3

```
*DIODE6.CIR–POWER SUPPLY
*** Description of transformer and bridge and filter ***
RZ 6 7 130
D5 0 7 D1N750
RLOAD 7 0 1K
.TRAN .016 .50 0 .60M
.PROBE
.END
```

The circuit description is straightforward, but special care should be taken to put the zener into the circuit with the correct polarity. The zener must be reverse biased. Node zero, the ground connection, at the anode of the zener, is the first node in the diode description.

The plots in Figure 4.8 show the voltage across the load lowered to approximately 4.74 Volts, and the peak-to-peak ripple voltage is 2 mV. This is a factor of 50 less than the 100 mV ripple in the previous unregulated circuit.

4.6 Clipper Circuits

A radio transmitter that broadcasts a frequency modulated (FM) signal outputs an AC waveform that has a constant peak amplitude. The information (music) is en-

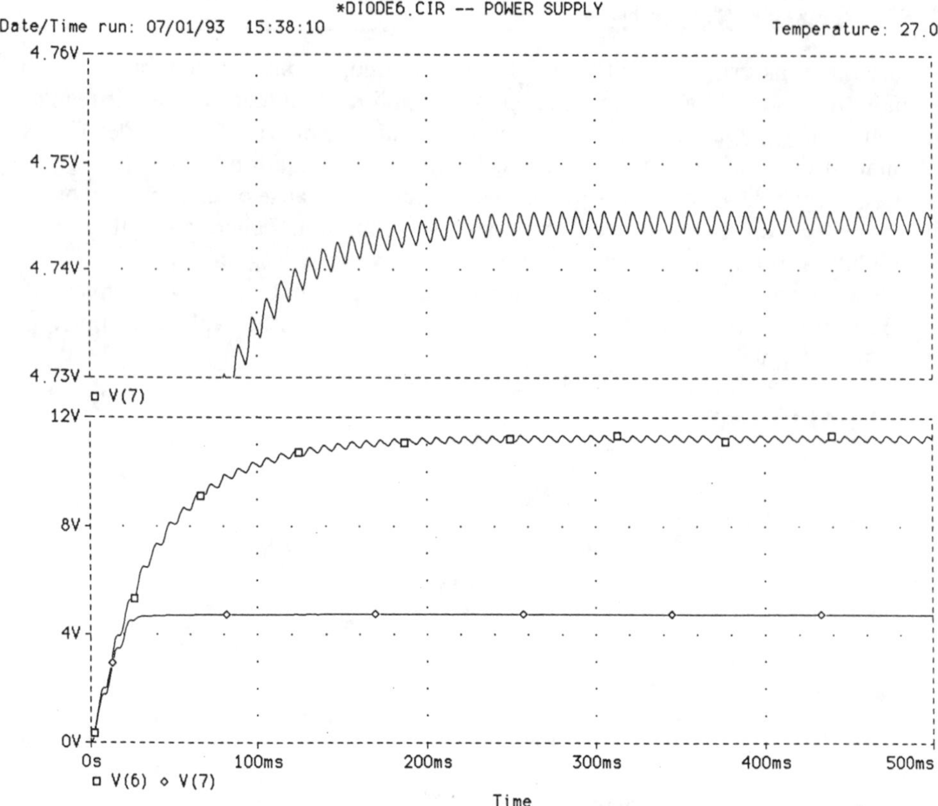

Figure 4.8

coded in variations of the carrier frequency, hence the name frequency modulation. FM receivers are designed to decode the frequency variations, but if they are to do this accurately, the signal must have a constant peak amplitude. The constant peak that existed at the transmitter can be lost to interference and noise in the front end of the receiver. To remake the signal into one of constant peak amplitude, clipper circuits are used.

While FM radios use transistors to accomplish the clipping effect, students usually first encounter the principle of clipping while studying zener diodes. The clipping of an AC wave can be effected by putting two zeners "nose-to-nose" in parallel with the load. Again the circuit description for the circuit in Example 4.4 is straightforward, but the nodes of the zeners must be listed in the correct orientation for the correct simulation to occur. Graphic output is shown in Figure 4.9.

```
*DIODE8.CIR–CLIPPER CIRCUIT
.INC \SPICE\DIODE.LIB
V1  1 0 SIN( 0 20 1000 )
R1  1 2 100
```

```
D1 2 3 D1N750
D2 0 3 D1N750
RLOAD 2 0 1K
.TRAN .05M 10M 0 10U
.PROBE
.END
```

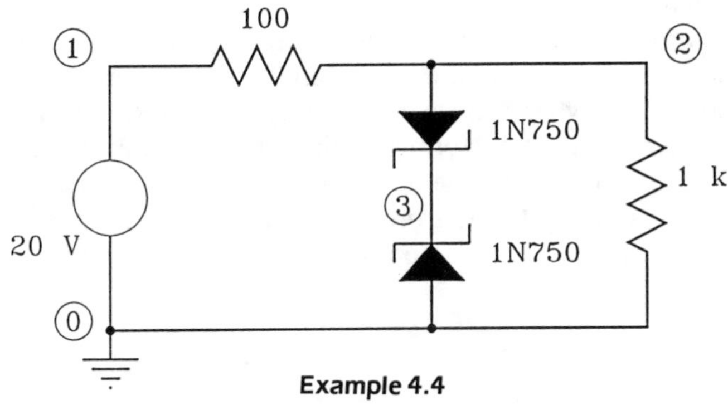

Example 4.4

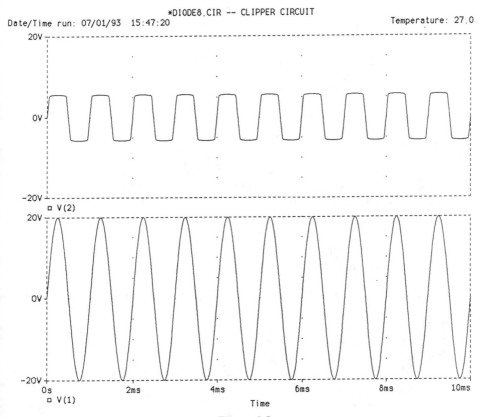

*DIODE8.CIR -- CLIPPER CIRCUIT
Date/Time run: 07/01/93 15:47:20
Temperature: 27.0
20V
0V
-20V
□ V(2)
20V
0V
-20V
0s 2ms 4ms 6ms 8ms 10ms
□ V(1)
Time

Figure 4.9

4.7 Subcircuits

A subcircuit is any portion of a circuit the user might find convenient to consider as a logical unit. This unit can be saved and easily inserted into any other circuit without rewriting all the individual lines of circuit description that make up the subcircuit. Subcircuits can be thought of as the PSpice analog to functions or subroutines in programming languages.

The concept of a subcircuit could be introduced anywhere in a book on PSpice, but it is the purpose of this text to present material on an "as needed" basis. Looking back at the power supply circuit, we might choose to consider the circuit as three separate units—namely, the transformer section, the bridge section, and the rest of the circuit. The transformer and the bridge sections represent functional units that can be rewritten as subcircuits and stored in a library. Diagramatically, we can explode the power supply schematic into three units, as shown in Figure 4.10.

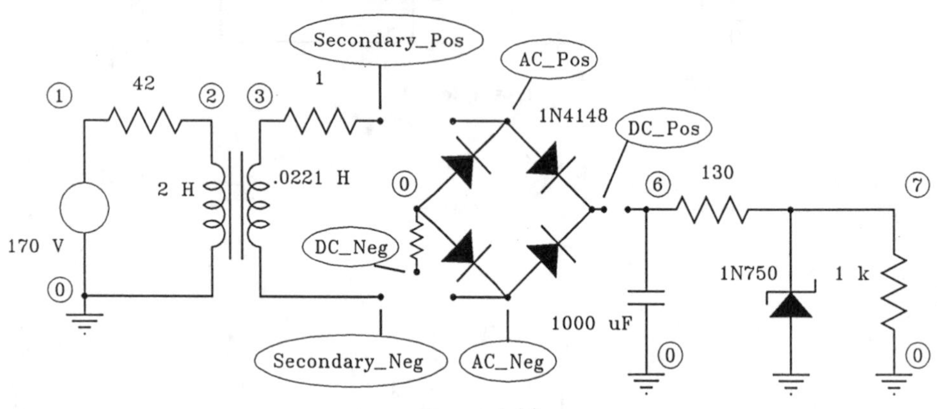

Figure 4.10

If the transformer section is written as a subcircuit and stored in the DIODE.LIB library, it can be inserted as a unit into any circuit description that includes DIODE. LIB. Instead of rewriting the same six lines of circuit description for every circuit powered by a 12.6 Volt transformer, a one-line subcircuit call can be used. The same is true for the bridge rectifier. Subcircuits are written in the same syntax as ordinary circuit descriptions but this requires a few extra rules. The transformer and bridge subcircuits are written as follows.

```
.SUBCKT 12V_TRANSFORMER  SECONDARY_POS SECONDARY_NEG
  V1 1 0 SIN( 0 170 60 )
  RPRIMARY     1 2 42
  LPRIMARY     2 0 2
  LSECONDARY   3 SECONDARY_NEG .0221
  K1           LPRIMARY LSECONDARY .999
  RSECONDARY   3 SECONDARY_POS 1
.ENDS
```

```
.SUBCKT BRIDGE AC_POS AC_NEG DC_POS DC_NEG
  D1 AC_POS DC_POS D1N4148
  D2 0      AC_POS D1N4148
  D3 0      AC_NEG D1N4148
  D4 AC_NEG DC_POS D1N4148
  R1 0      DC_NEG .001
.ENDS
```

The general form for the subcircuit command is

.SUBCKT <subcircuit_name> <connecting_nodes>

Several special rules apply to the description of subcircuits.

1. Subcircuits begin with the .SUBCKT line and end with .ENDS.
2. The connecting nodes for a subcircuit are named with *words* instead of *numbers*. This is not strictly necessary, but it is a good idea. Using descriptive node names for the BRIDGE subcircuit makes it obvious which nodes should connect to the AC signal from the previous stage and which nodes output the DC voltage for the following stage.
3. Each subcircuit must have a grounded node (node zero).
4. Node zero may not be used as a connecting node. Notice in the BRIDGE subcircuit, a small (.001 ohm) resistor has been added. This is done because a bridge has four nodes, and all four are connecting nodes (i.e., none can be named zero). The addition of this fifth component creates a fifth, nonconnecting node that can be named zero.
5. The names of subcircuit components and nodes are local to the subcircuit. PSpice will not confuse a node named "1" in a subcircuit with node "1" in the main circuit description. Similarly a resistor "R1" can be described in a subcircuit and the same name reused in the main circuit.

Subcircuits are included in circuit description using the symbol "X<name>." The general form is

X<name> <connecting_nodes> <subcircuit_name>

The order of the connecting node in the circuit description must correspond to the order of the nodes in the subcircuit description. DIODE6.CIR has been rewritten using the two subcircuits that have been added to DIODE.LIB. Notice that in the bridge description, node "1" corresponds to AC_POS, "2" corresponds to AC_NEG, and so on.

```
*DIODE6.CIR—POWER SUPPLY
.INC \SPICE\DIODE.LIB
```

```
X1 1 2 12V_TRANSFORMER
X2 1 2 3 0 BRIDGE
C1 3 0 1000U
RZ 3 7 130
D5 0 7 D1N750
RLOAD 7 0 1K
.TRAN .016 .20 0 .60M
.PROBE
.END
```

5

Operational Amplifier Circuits

Opinions vary about which topic should follow the study of diode circuits. Some curriculums proceed to transistor theory, on the grounds that the progression from one-junction to two-junction devices is the next logical step. Other schools choose to study operational amplifiers (op amps) because of their "black box" simplicity. Op amps allow the examination of amplifiers, oscillators, and the like without introducing the difficulty of biasing the transistors. For classroom and laboratory work using PSpice, the order in which the topics are studied is unimportant. The chapters on the use of PSpice in op amp and in transistor circuits are self-contained, and their order can be interchanged without losing continuity.

5.1 Creating OPAMP.LIB

As is the case with diodes, there is no single accurate model for an op amp. Real op amps differ greatly in such characteristics as input impedance, bandwidth, output current, and offset voltage and current. Consequently, there is a need for a library of models for specific op amps. If the choice is made to use the .INC statement to include the diode models, it is necessary to create OPAMP.LIB. The advantage to using the .INC statement is the inclusion of the diode models in the .OUT file. Although many diode parameters are of little use to the user, having the connecting nodes for the model near the circuit description can be handy when trouble shooting the circuit. The EVAL.LIB supplied with the evaluation version of PSpice contains models for the LM324 and the UA741 op amps. To make OPAMP.LIB use the DOS copy command to make a copy of EVAL.LIB called OPAMP.LIB and then use an editor to delete everything except the models for the two op amps. Be sure to copy the comment lines preceding the subcircuit model. These lines contain needed information concerning the order for the connecting nodes in the main circuit.

The new library should contain the following information. Notice that "**** more data ****" has been substituted for the bulk of the subcircuit component descriptions to conserve space here. **This should not be done in the OPAMP.LIB library.**

```
* connections:      non-inverting input
*                   | inverting input
*                   | | positive power supply
*                   | | | negative power supply
*                   | | | | output
*                   | | | | |
.subckt LM324       1 2 3 4 5
*
  c1      11 12 2.887E-12
  c2       6  7 30.00E-12
*
**** more data ****
*
  vln     0 92 dc 40
.model dx D(Is=800.0E-18 Rs=1)
.model qx PNP(Is=800.0E-18 Bf=166.7)
.ends

*----------------------------------------------------------------------

* connections:      non-inverting input
*                   | inverting input
*                   | | positive power supply
*                   | | | negative power supply
*                   | | | | output
*                   | | | | |
.subckt uA741       1 2 3 4 5
*
  c1      11 12 8.661E-12
  c2       6  7 30.00E-12
*
**** more data ****
*
  vln     0 92 dc 40
.model dx D(Is=800.0E-18 Rs=1)
.model qx NPN(Is=800.0E-18 Bf=93.75)
.ends
```

5.2 Level Detector Circuit

At its simplest, an op amp can be used without feedback to generate a pulse whenever the input exceeds a preset voltage level. The circuit shown in Example 5.1 uses the high open loop gain of the op amp to drive the output to near saturation levels. The output is positive when the AC voltage is higher than the DC level on the inverting input, and negative otherwise.

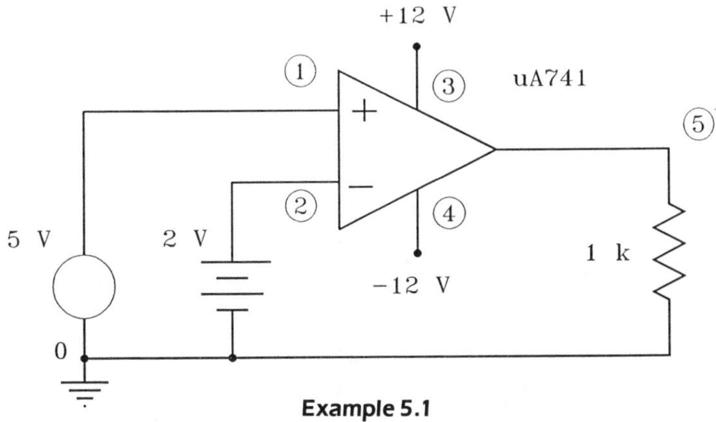

Example 5.1

```
*OPAMP1.CIR—LEVEL DETECTOR
.INC \SPICE\OPAMP.LIB
V1 3 0 12
V2 4 0 −12
V3 1 0 SIN(0 5 100)
V4 2 0 2
RLOAD 5 0 1K
X1 1 2 3 4 5 UA741
.TRAN 1M 50M 0 .1M
.PROBE
.END
```

The op amp is included in a circuit description as a subcircuit. This necessitates the inclusion of the OPAMP.LIB library. In this case "X1" is the call to the subcircuit for a UA741 op amp. The UA741 has five connecting nodes. In this circuit, the connecting nodes have the same numbers as the nodes in the subcircuit description in the library. In general this is not the case, as is seen in subsequent examples. The op amp subcircuit is used in exactly the same manner as the subcircuit for the bridge rectifier described in Chapter 4. It is critically important to list the connecting nodes in the correct order. This is the reason for keeping the connection comment lines in the library as a reference, even though the subcircuit would work if they were deleted.

A second item to note in the circuit description is the description for the negative supply for the op amp. "V2" is considered to have node 4 for its positive end, but the supply itself is a negative voltage. It might be clearer to describe "V2" as a positive supply with the plus end connected to ground. That is

```
V2 0 4 12
```

Either way, the circuit will work correctly giving the output shown in Figure 5.1.

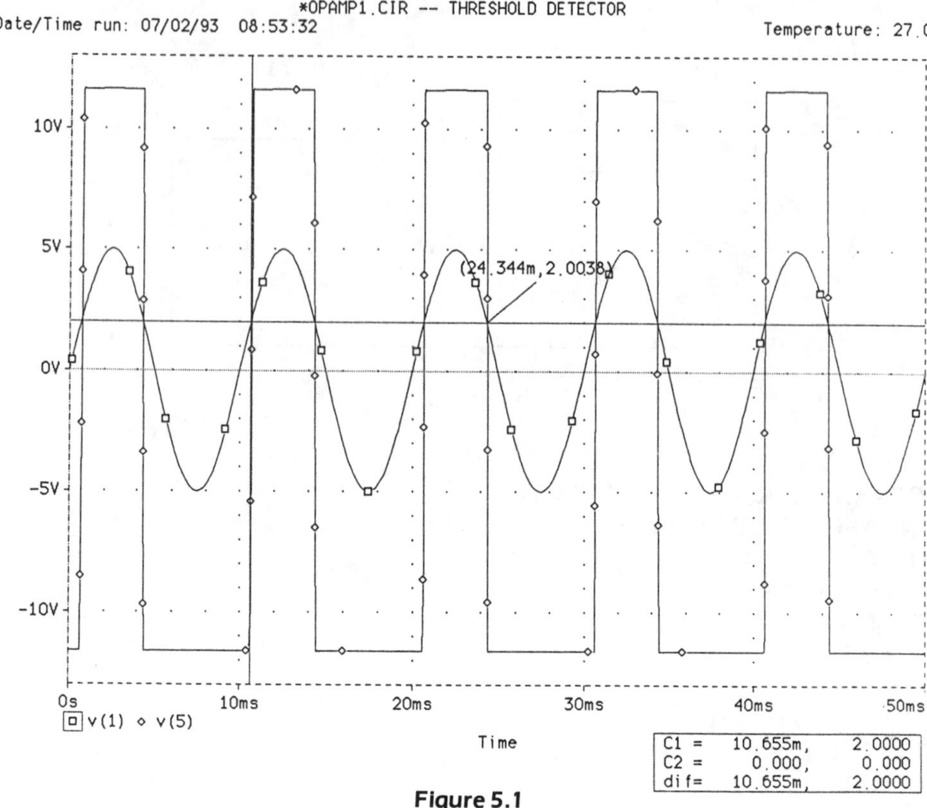

Figure 5.1

5.3 Inverting Amplifier

Op amps used as signal amplifiers provide a simple way to add significant gain to low-frequency signals. The following amplifier is analyzed twice. The first analysis is performed on the circuit description exactly as it is written. For the second analysis, the asterisks are moved from the AC source and the .AC analysis lines to the sine wave source and the transient analysis lines. This provides gain limit information for any given frequency input wave and for the unity gain bandwidth. The amplifier is configured as an inverting amplifier, as shown in Example 5.2. The circuit description is straightforward. The singular item to note is the connecting nodes for the op amp subcircuit. For reference, the connection information is repeated here.

```
* connections:   non-inverting input
*                | inverting input
*                | | positive power supply
*                | | | negative power supply
*                | | | | output
*                | | | | |
*.subckt uA741   1 2 3 4 5
```

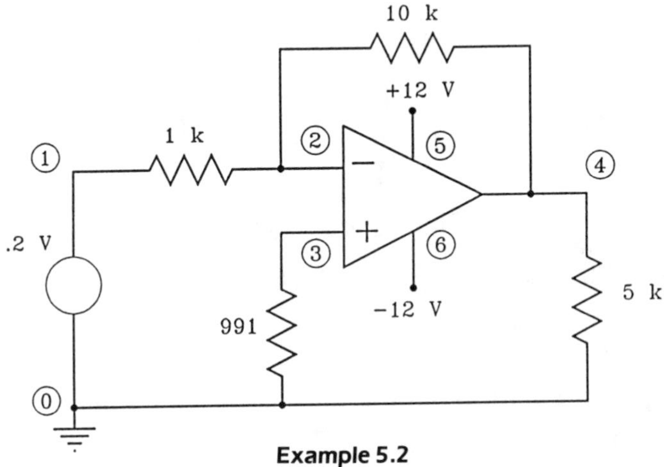

Example 5.2

```
*OPAMP2.CIR—INVERTING AMPLIFIER
.INC \SPICE\OPAMP.LIB
V1 5 0 12
V2 0 6 12
V3 1 0 SIN( 0 .02 100)
*V3 1 0 AC .02
RIN 1 2 1K
RFEEDBACK 2 4 100K
RCOMP 3 0 991
RLOAD 4 0 5K
X1 3 2 5 6 4 UA741
.TRAN 1M 50M 0 .1M
*.AC DEC 100 10 10MEG
.PROBE
.END
```

The graphic output given by PROBE (Figure 5.2) shows the output is 100 times larger than the input voltage, as expected.

Altering the circuit for a frequency-based analysis by using the alternate V3 source and changing from transient analysis to the .AC frequency-based analysis yields the results shown in Figure 5.3.

PROBE makes clear two expected features. First, the unity gain bandwidth is approximately 871 kHz. And second, labeling two points one decade apart shows that the gain decreases at the expected rate of 20 dB.

5.4 Variable-Gain Amplifiers—Modeling Variable Resistors

Making the gain variable in an op amp circuit is a simple matter of replacing the fixed feedback resistor with a variable resistor. PSpice has the ability to model a

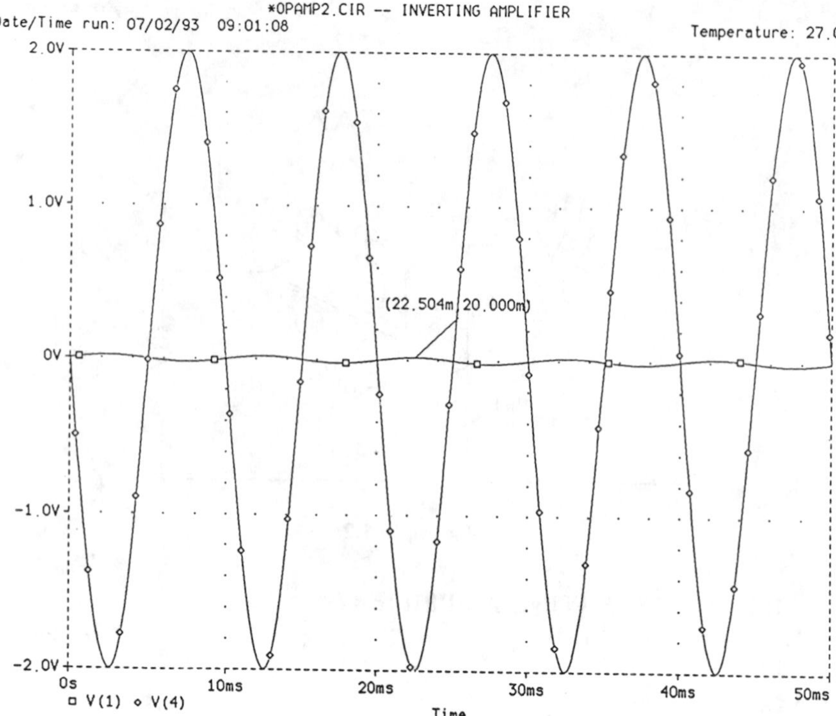

Figure 5.2

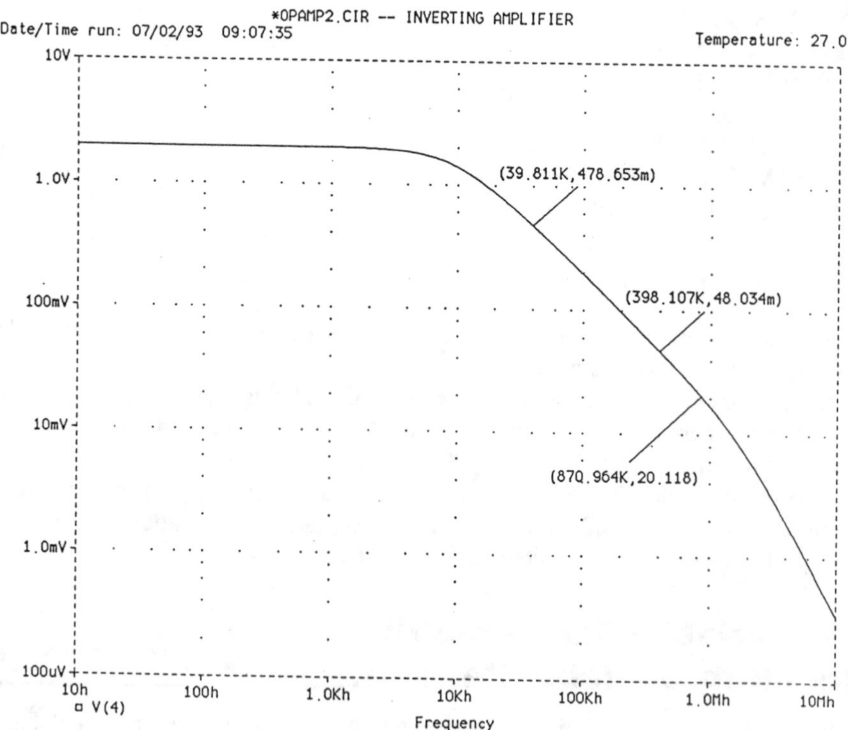

Figure 5.3

resistor that varies over a definite range by definite increments. Since a varying graph would be difficult for the user to read (and difficult for the writers of PSpice to produce), it is more reasonable to have PSpice put values in the .OUT file for each increment of the variable resistor. Making the feedback resistor variable, the circuit becomes

```
*OPAMP3.CIR—VARIABLE GAIN AMPLIFIER
.INC \SPICE\OPAMP.LIB
V1 5 0 12
V2 0 6 12
V3 1 0 AC .02
RIN 1 2 1K
.PARAM VALUE = 1K
RFEEDBACK 2 4 {VALUE}
.STEP PARAM VALUE 10K 100K 10K
RCOMP 3 0 991
RLOAD 4 0 5K
X1 3 2 5 6 4 UA741
.AC DEC 1 100 100
.PRINT AC V(1) V(4)
.END
```

The three lines beginning with .PARAM VALUE = 1K are used to indicate that a variable resistor is used between nodes 2 and 4. .PARAM creates a variable called VALUE. A variable must be given an initial value, so VALUE is set equal to 1K. In this circuit, the initial value is irrelevant because it will be overwritten by the .STEP PARAM statement. Next, the feedback resistor is set equal to VALUE. Unlike fixed values, variables must be enclosed in curly braces. Without the statement to STEP, the variable VALUE from 10K to 100K in 10K increments, the circuit would be analyzed using a feedback resistor of 1K (i.e., the value of VALUE). With the STEP statement, the analysis is performed 10 times for 10 different values of the feedback resistor. The resulting values for the output voltage, V(4), are printed to the .OUT file. Unfortunately, the data are not conveniently tabled, and some editing of the .OUT file is required to bring the data into a neat and readable form.

FREQ	V(1)	V(4)
1.000E+02	2.000E−02	2.000E−01
1.000E+02	2.000E−02	4.000E−01
1.000E+02	2.000E−02	5.999E−01
1.000E+02	2.000E−02	7.998E−01
1.000E+02	2.000E−02	9.997E−01
1.000E+02	2.000E−02	1.200E+00
1.000E+02	2.000E−02	1.399E+00
1.000E+02	2.000E−02	1.599E+00
1.000E+02	2.000E−02	1.799E+00
1.000E+02	2.000E−02	1.999E+00

5.5 Potentiometer Subcircuit

Although it is not difficult to place a variable resistor in a circuit description, it can be more convenient to model a potentiometer as a subcircuit and save it in a library. The subcircuit description given here and shown in Figure 5.4 applies to a true three-terminal potentiometer rather than the two-terminal variable resistor used in Example 5.2.

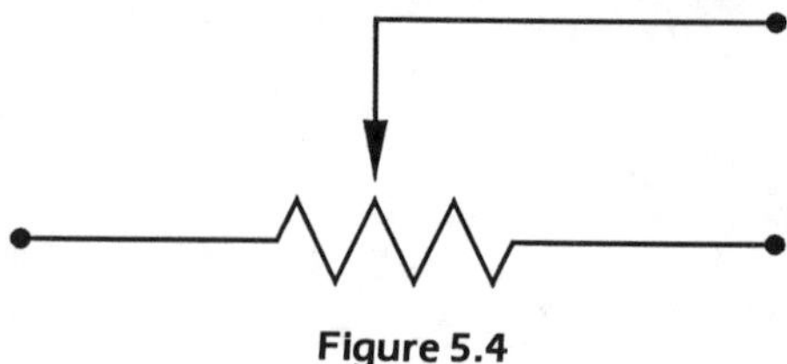

Figure 5.4

```
.subckt pot (oneend, center, otherend) params: r=1 set=.1
rleft oneend center {(1.001 – set)*r}
rright center otherend { (.001 + set)*r}
.ends
```

This describes a three-terminal device with a total end-to-end resistance of "r" and position setting for the wiper given by "set." "r" and "set" both require initial values, but these values are irrelevant and will be overwritten by values supplied by the main circuit description. "rleft" is a resistor named "left" with nodes "oneend" and "center." The value of "rleft" is given by a formula rather than a fixed number. This allows the main circuit description to pass a different value for "set" for each of 10 different analyses. Notice that a component value given by a formula must be enclosed in curly braces.

If this model is added at the end of OPAMP.LIB, the pot can be called as a subcircuit. To use this subcircuit, OP3.CIR must be altered by removing the three lines that describe the variable resistor:

```
*.PARAM VALUE = 1K
*RFEEDBACK 2 4 {VALUE}
*.STEP PARAM VALUE 10K 100K 10K
```

and replacing them with

```
.PARAM SET = 5.
.STEP PARAM(SET) .1 1 .1
XPOT 2 4 4 POT PARAMS: R = 100K SET = {SET}
```

On the surface, replacing three lines with three alternative lines seems to be not worthwhile. The advantage gained is the availability of the second section of the pot. In circuit OP3.CIR, only the resistance between one end and the wiper is needed, so the end node of the second section is tied to the wiper by the defined nodes of the pot

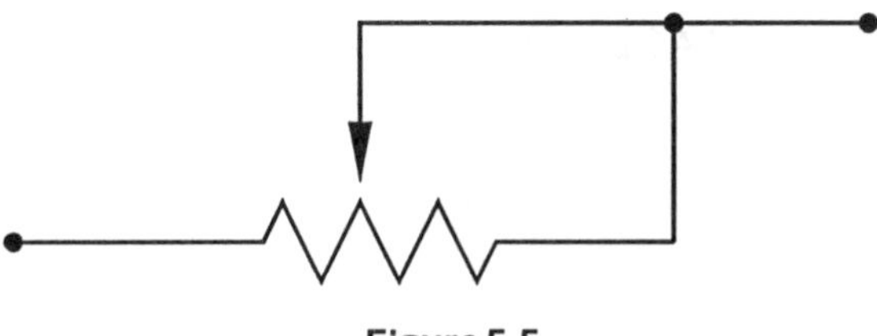

Figure 5.5

(XPOT 2 4 4). Because PSpice will not accept a floating node, it must be tied to the output node, effectively shorting the second section, as shown in Figure 5.5.

The .PARAM statement creates the variable SET and gives it the required temporary value of 5. The .STEP statement instructs PSpice to step the variable "SET" through the values from .1 to 1 in .1 increments. The XPOT line gives the nodes for the pot, the model name, and the values to be passed to the subcircuit. The general form for the .STEP statement is

.STEP <sweep_variable> <initial_value> <final_value> <increment>

5.6 Multistage Amplifiers

The analysis of multistage amplifiers is within the capabilities of PSpice. The amplifier in Example 5.3 has been limited to two stages for the benefit of readers using old versions of PSpice. Versions 5.1 and 5.4 were confined to conventional memory and would produce an error message if a circuit contained a third op amp. Version 6.0 no longer has the two op amp limit, and more stages can be added to the circuit description and simulated. Example 5.3 uses an inverting amplifier for the first stage and a noninverting second stage. The output from PROBE shows the expected gain of two in the first stage and three in the second. No new PSpice features are included in this example. Consequently the circuit (Example 5.3), circuit description, and the analysis output (Figure 5.6) are provided without further comment.

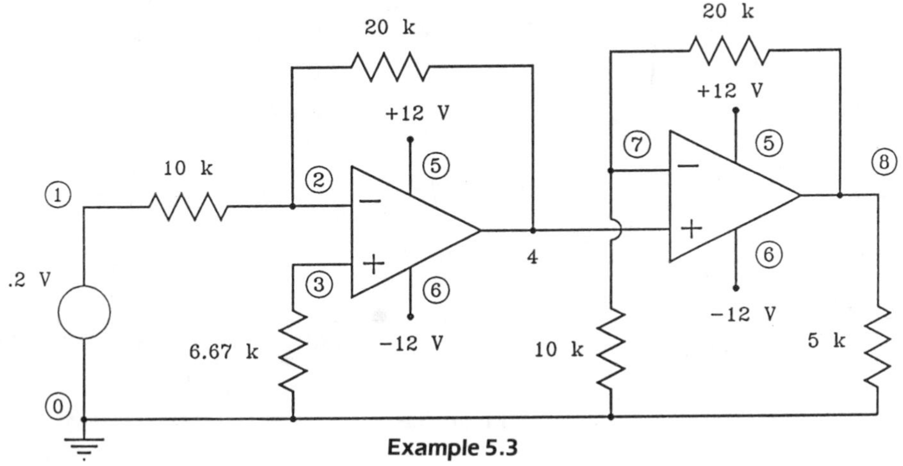

Example 5.3

```
*OPAMP4.CIR–TWO STAGE AMP
.INC \SPICE\OPAMP.LIB
V1 5 0 12
V2 0 6 12
V3 1 0 SIN(0 .2 100)
RIN1 1 2 10K
RCOMP 3 0 6.67K
RFEEDBACK1 2 4 20K
RFEEDBACK2 8 7 20K
RIN2 7 0 10K
RLOAD 8 0 5K
X1 3 2 5 6 4 UA741
X2 4 7 5 6 8 UA741
.TRAN 1M 50M 0 .1M
.PROBE
.END
```

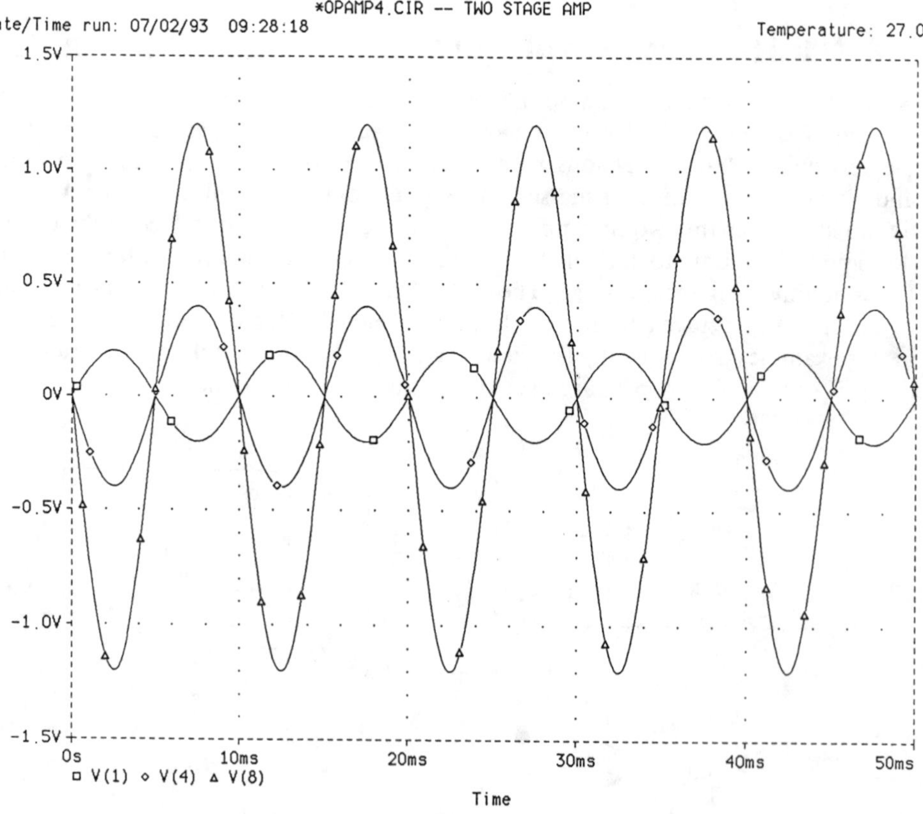

Figure 5.6

5.7 Op Amp Oscillators

Simulated oscillators are much like real oscillators in that they sometimes refuse to oscillate. In the laboratory, the problem is usually the result of an overlooked design error. In simulation, the problem is often traceable to the "perfect" nature of the simulated components. Real oscillators often rely on minuscule ever-present noise or the shock of a sudden power-up, to jump-start the oscillations. In simulation, this noise is missing, and students find that a circuit that oscillates perfectly in the lab refuses to oscillate in simulation. While some designs stubbornly refuse to oscillate, others break into oscillation only after a quiet initial interval. If the simulation is not observed for a long enough time, the circuit appears not to perform.

One answer to this problem is to include a small noise source in the circuit. PSpice can model a noise source, but such a model is a little messy and there is a simpler solution.

The 100 Hz oscillator design shown in Example 5.4 was chosen because it begins to oscillate only after approximately 80 milliseconds. If the transient analysis is performed for only 50 milliseconds, the expected five cycles do not appear, and the oscillator seems to not function.

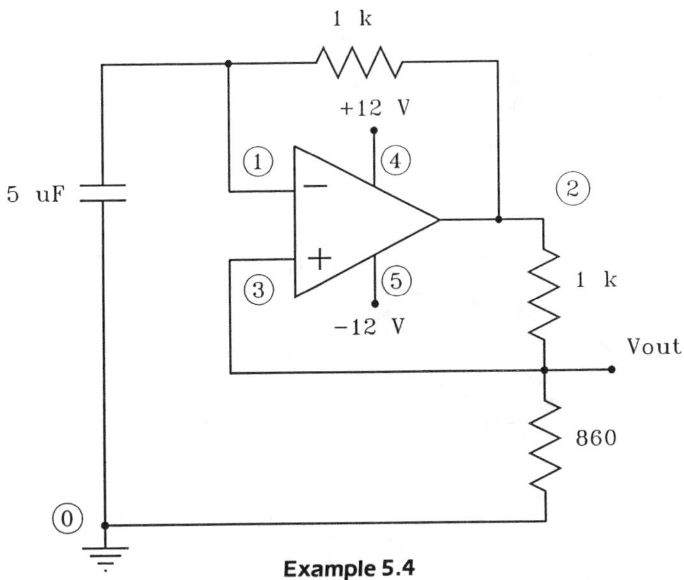

Example 5.4

```
*OPAMP5.CIR—OSCILLATOR CIRCUIT
.INC \SPICE\OPAMP.LIB
*V1 4 0 PULSE( 0 12 0 1N 1N 100 101 )
*V2 0 5 PULSE( 0 12 0 1N 1N 100 101 )
V1 4 0 12
V2 0 5 12
RFEEDBACK 2 1 1K
```

```
C1 1 0 5U
R1 2 3 1K
ROUT 3 0 860
X1 3 1 4 5 2 LM324
.TRAN .1M 200
.PROBE
.END
```

The two supplies powering the op amp are described twice. For the first analysis, V1 and V2 are described as DC supplies as in previous examples. This results in a delayed oscillation, as shown in Figure 5.7.

The alternative to using DC supplies for the op amp is to describe V1 and V2 as pulsing supplies that quickly rise to 12 volts and remain there until long after the simulation is complete. In this case, the pulse stays at 12 Volts for 100 seconds, long enough for 10,000 cycles. The advantage to using the PULSE statement is the 1 nanosecond rise time. This quick voltage change nudges the circuit into immediate oscillation. The revised circuit begins to oscillate immediately and produces the output shown in Figure 5.8.

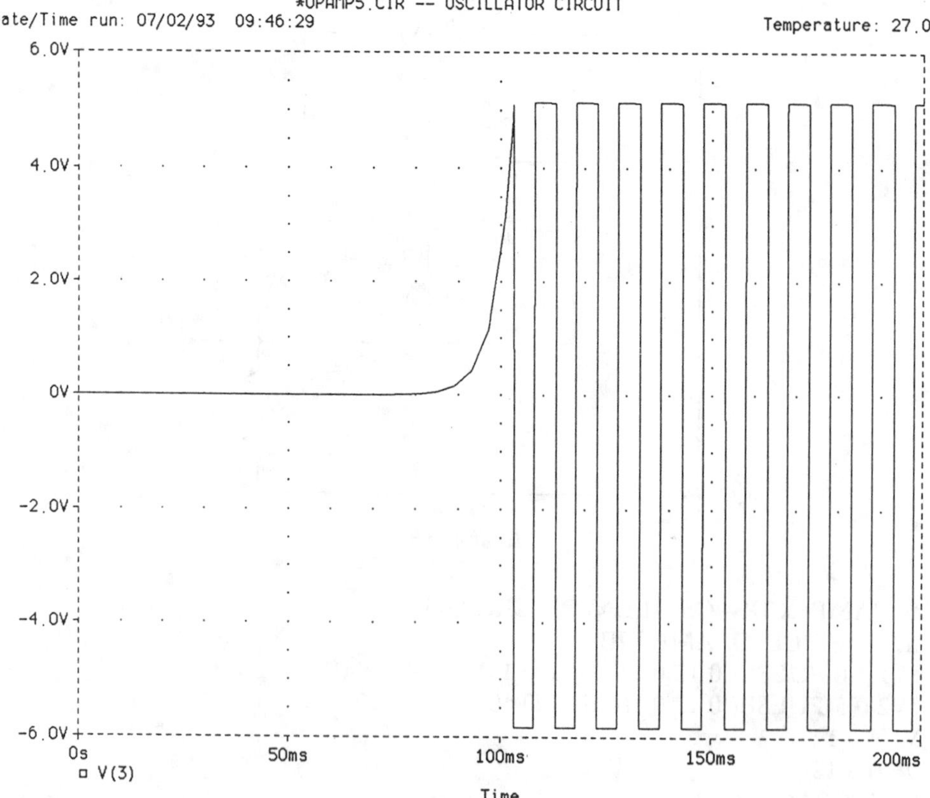

Figure 5.7

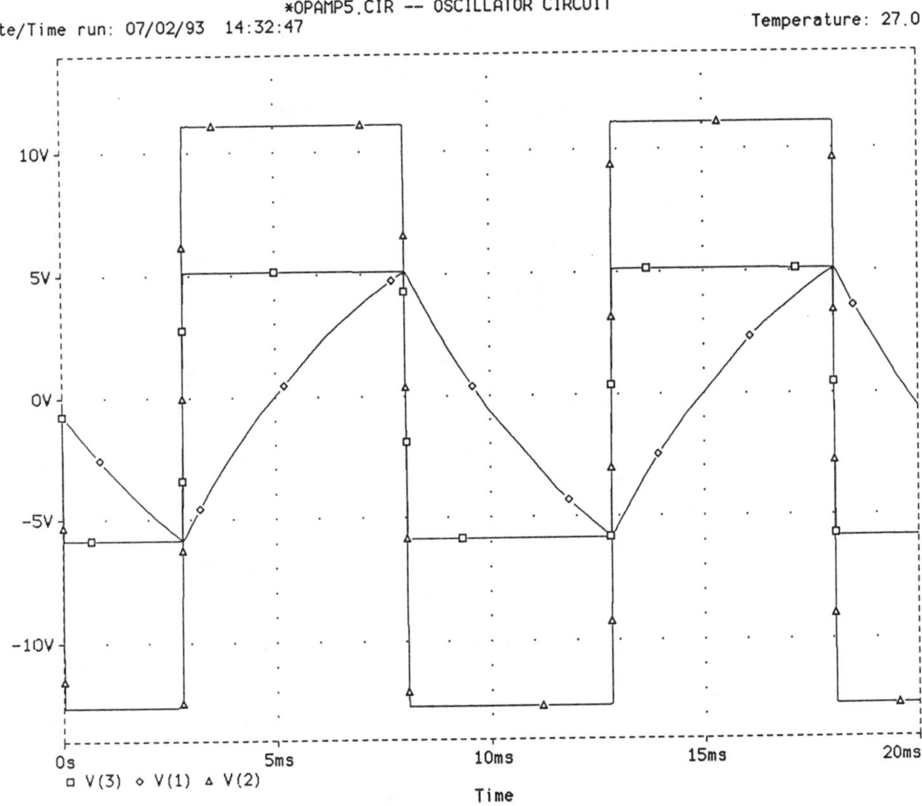

Figure 5.8

Op amps provide a simple means for designing oscillators if the required oscillating frequency is not too high. If the capacitor in the previous circuit description is made 100 times smaller, the frequency theoretically should be 100 times higher. However, the op amp requires a finite amount of time to respond to the feedback voltage. The result is the loss of the clean square wave produced at 100 Hz and a deviation from the expected 10 kHz down to about 6 kHz, as shown in Figure 5.9.

5.8 Precision Half-Wave Rectifier

Through use of an op amp, a half-wave rectifier can be designed with characteristics superior to those of the single-diode rectifier described in Chapter 4. The circuit shown in Example 5.5 also illustrates the inclusion of two libraries in a single circuit description.

```
*OPAMP6.CIR—PRECISION HALF-WAVE RECTIFIER
.INC \SPICE\DIODE.LIB
.INC \SPICE\OPAMP.LIB
```

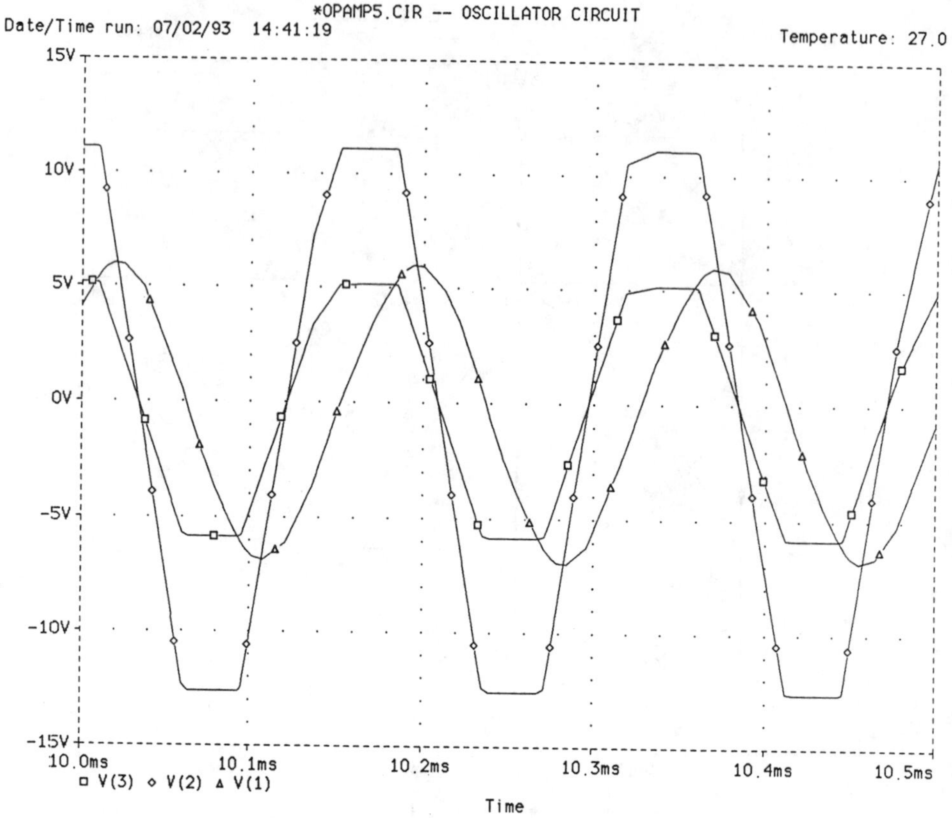

*OPAMP5.CIR -- OSCILLATOR CIRCUIT
Date/Time run: 07/02/93 14:41:19
Temperature: 27.0
15V
10V
5V
0V
-5V
-10V
-15V
10.0ms
10.1ms
10.2ms
10.3ms
10.4ms
10.5ms
□ V(3) ◇ V(2) Δ V(1)
Time

Figure 5.9

1 k
1N4148
+15 V
1 k
1
2
7
1N4148
8
6
2 V
4
5 k
−15 V
0

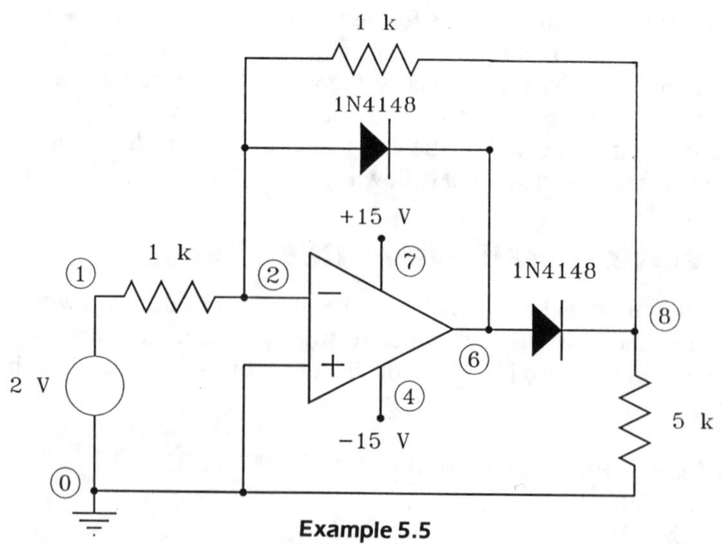

Example 5.5

```
V1 7 0 15
V2 0 4 15
V3 1 0 SIN(0 2 60)
RIN 1 2 1K
RFEEDBACK 2 6 1K
D1 2 6 D1N4148
D2 6 8 D1N4148
X1 0 2 7 4 6 UA741
RLOAD 8 0 5K
.TRAN 10M 80M 0 .3M
.PROBE
.END
```

The output for this circuit, graphed in Figure 5.10, is clearly missing the distortion and voltage drop present in the output of the simpler single-diode rectifier because of the nonlinear characteristics of the diode.

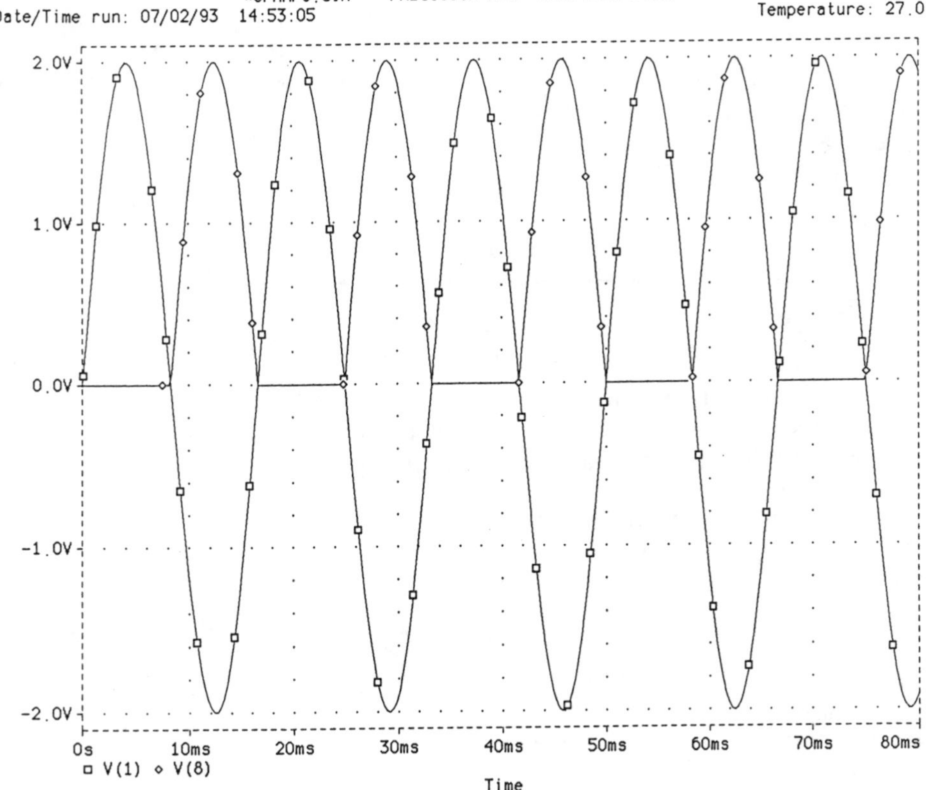

Figure 5.10

5.9 60 dB Active Filter

Active filters have superior characteristics compared to passive filters. Example 5.6 is a two-stage Butterworth filter. Again, the circuit has been limited to two op amps for the benefit of users of versions 5.1 and 5.4. Users of version 6.0 can add more stages for the greater high-frequency suppression. This particular filter was chosen for examination because it illustrates the accuracy of PSpice simulations. The component values were chosen to give the filter a cutoff frequency of 15 kHz and a 60 dB rolloff.

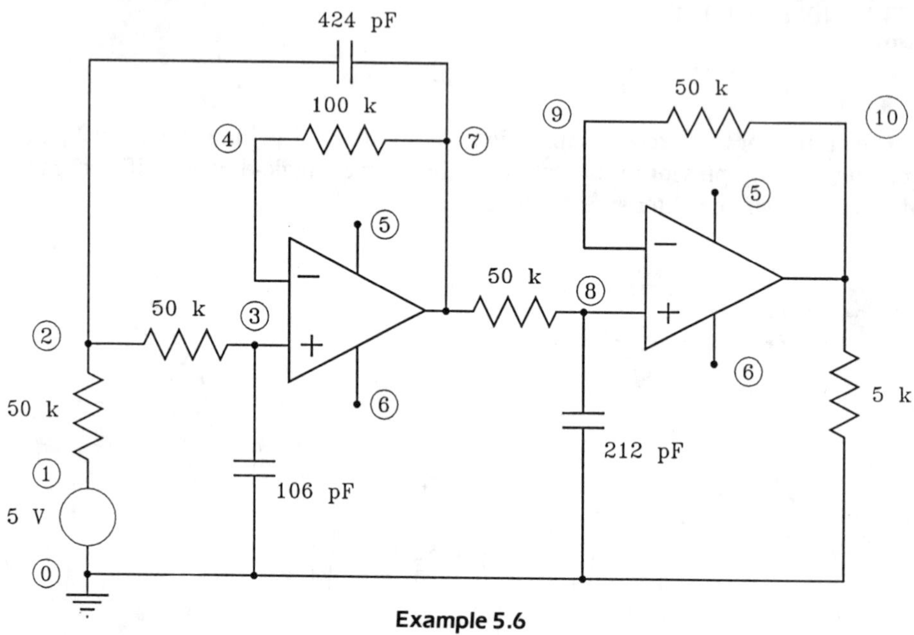

Example 5.6

The circuit is analyzed twice, first using an AC source for a frequency-based analysis, and second using a 15 kHz square wave source to illustrate its severe harmonic suppression. The frequency-based circuit description is

```
*OPAMP7.CIR—60dB ACTIVE FILTER
.INC \SPICE\OPAMP.LIB
V1 5 0 15
V2 0 6 15
V3 1 0 AC 5
R1 1 2 50K
R2 2 3 50K
R3 4 7 100K
R4 7 8 50K
R5 9 10 50K
RLOAD 10 0 5K
C1 2 7 424P
```

```
C2 3 0 106P
C3 8 0 212P
X1 3 4 5 6 7 UA741
X2 8 9 5 6 10 UA741
.AC DEC 100 10 10MEG
.PROBE
.END
```

The graphic output for the circuit, shown in Figure 5.11, is developed by first plotting the output voltage at node 10 and then using the "Plot_control" to add a second plot to plot the output voltage a second time. It is then necessary to return to the main menu and choose "X_axis," to set the X-axis range from 1K to 10M (remember megahertz is symbolized by "M" when using PROBE). The scale for each Y-axis has been changed to a logarithmic scale and the range changed from 1 V to 10 V for the upper plot only. Once the scale changes have been made, the three points of interest are more easily examined. Choosing "Cursor" at the main menu and moving the cross hairs along the graph using the mouse and the arrow keys permits the user

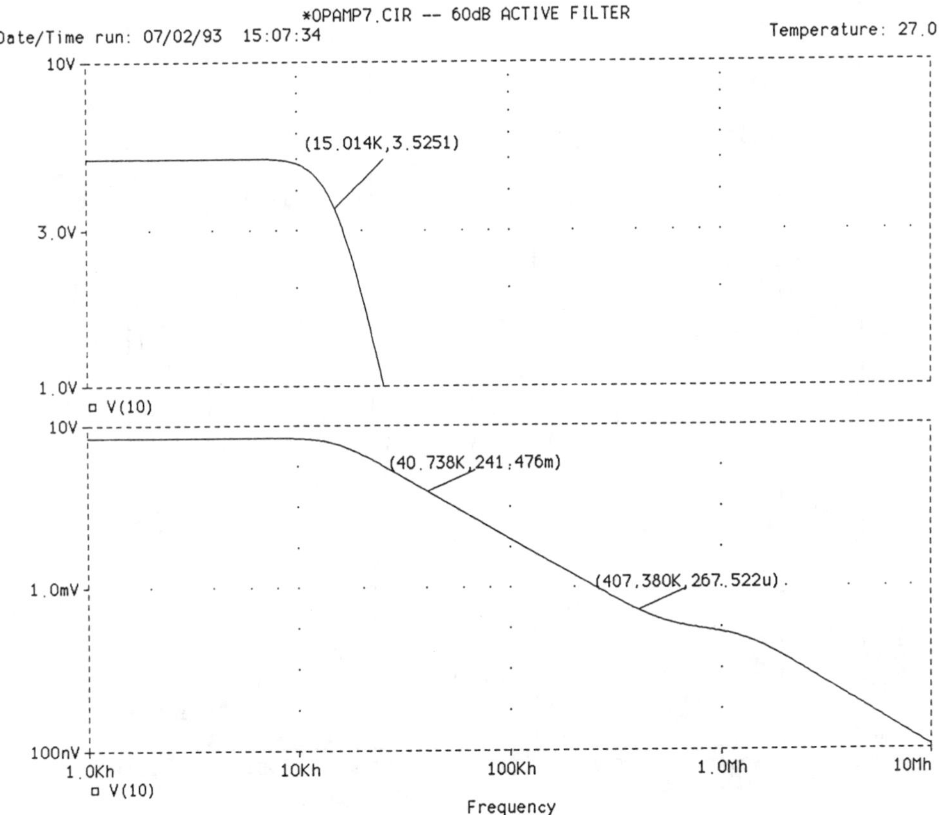

Figure 5.11

to locate the 15 kHz point by watching the coordinates of C1 (Cursor one) at the bottom right of the screen. Where the 15 kHz point is labeled, the voltage is found to be 3.525 Volts. This is close to the expected value of 3.535, or 5 Volts multiplied by .707. In other words, at the designed cutoff frequency of 15 kHz, the output voltage is suppressed by 3 dB.

A return to the main menu and the use of "Plot_control," permits the lower plot to be selected. On this plot, two points were selected one decade apart, both on the linear portion of the graph. Calculating the decibel suppression using the corresponding Y-axis values yields a rolloff slightly greater than 59.1 dB.

Equally interesting results are obtained from the time-based analysis. For this analysis, both the AC input described for V3 and the .AC analysis need to be changed to

```
V3 1 0 PULSE(0 5 0 1N 1N 33.33U 66.67U)
.TRAN 33U 333U 0 1U
```

The pulse is described as having a high level of 5 Volts, a 50% duty cycle, and a period of 66.67 microseconds—i.e., a frequency of 15 kHz.

The graphic analysis in Figure 5.12 shows that the square wave at the input resulted

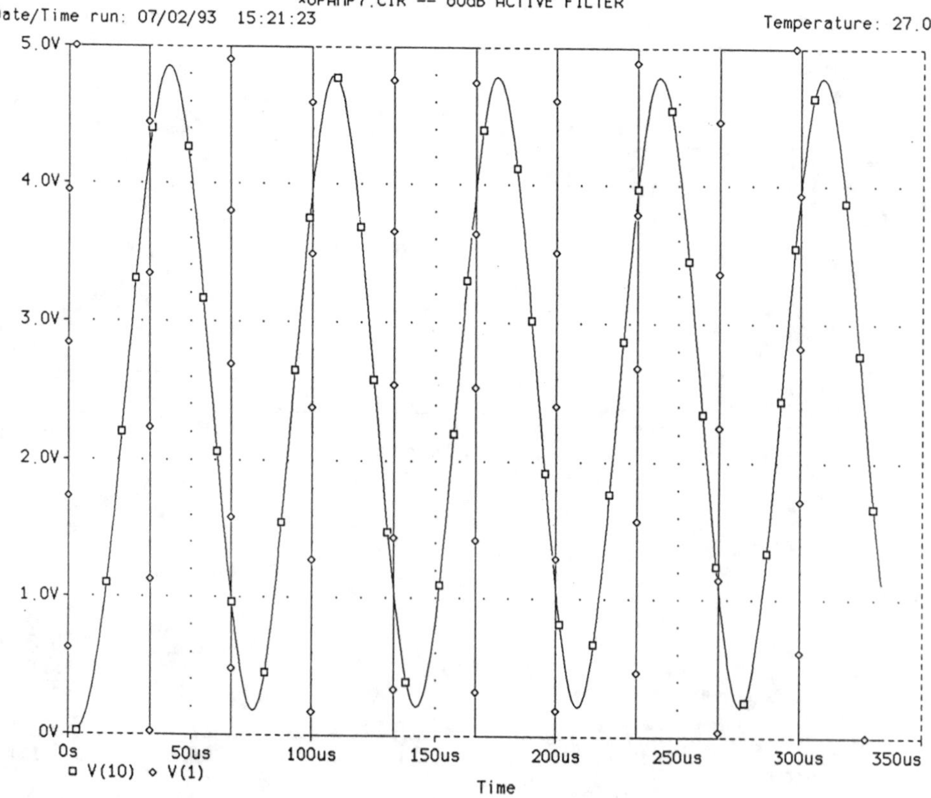

Figure 5.12

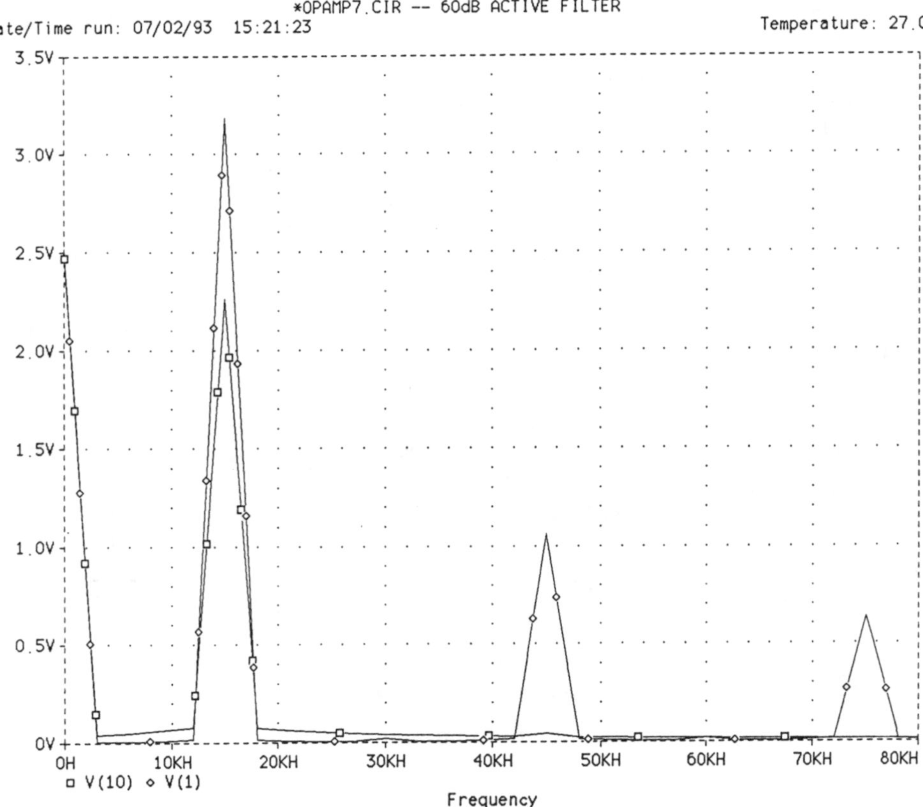

Figure 5.13

in an apparently undistorted sine wave at the output. In other words, the third, fifth, seventh, and so on harmonics known to be present in the square wave seem to have been removed. Fortunately it is not necessary to guess about the harmonic suppression characteristics of the circuit. Choosing "X_axis" and "Fourier" and adjusting the range of the X-axis produces the graph shown in Figure 5.13. This plot makes it clear that the third harmonic is greatly reduced and the higher harmonics have been eliminated.

5.10 Differential Amplifier

During their studies of operational amplifiers, students are often assigned the lab exercise of building an instrumentation amplifier. Instrumentation amplifiers require three op amps, and can be modeled by version 6.0 but not by earlier versions of PSpice. However, by eliminating the final op amp, a differential amplifier can be modeled by all versions, and 6.0 users can easily add the final stage. Example 5.7 has been designed to amplify the difference between the two inputs by a factor of five. At the peak of the sine wave inputs, the difference is .01 Volts, producing an output of .05 Volts.

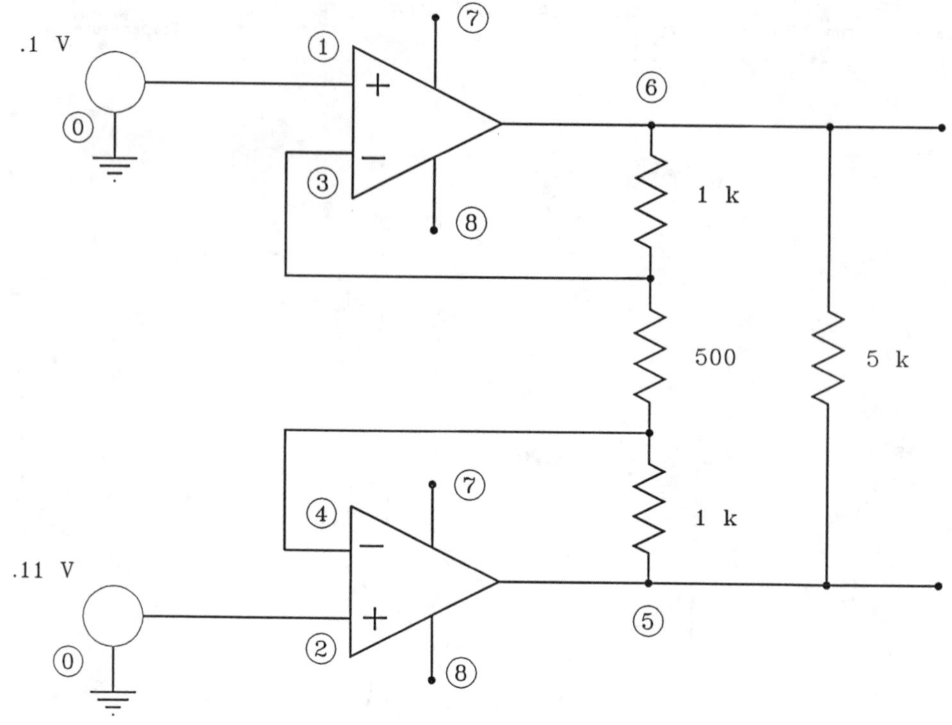

Example 5.7

```
*OPAMP8.CIR—DIFFERENTIAL AMPLIFIER
.INC \SPICE\OPAMP.LIB
V1 7 0 12
V2 0 8 12
V3 1 0 SIN( 0 .1 100 )
V4 2 0 SIN( 0 .11 100 )
R1 6 3 1K
R2 3 4 500
R3 4 5 1K
RLOAD 6 5 5K
X1 1 3 7 8 6 LM324
X2 2 4 7 8 5 LM324
.TRAN 1M 50M 0 .1M
.PROBE
.END
```

Since no new features are introduced in the circuit description, a few bells and whistles are added to the output.

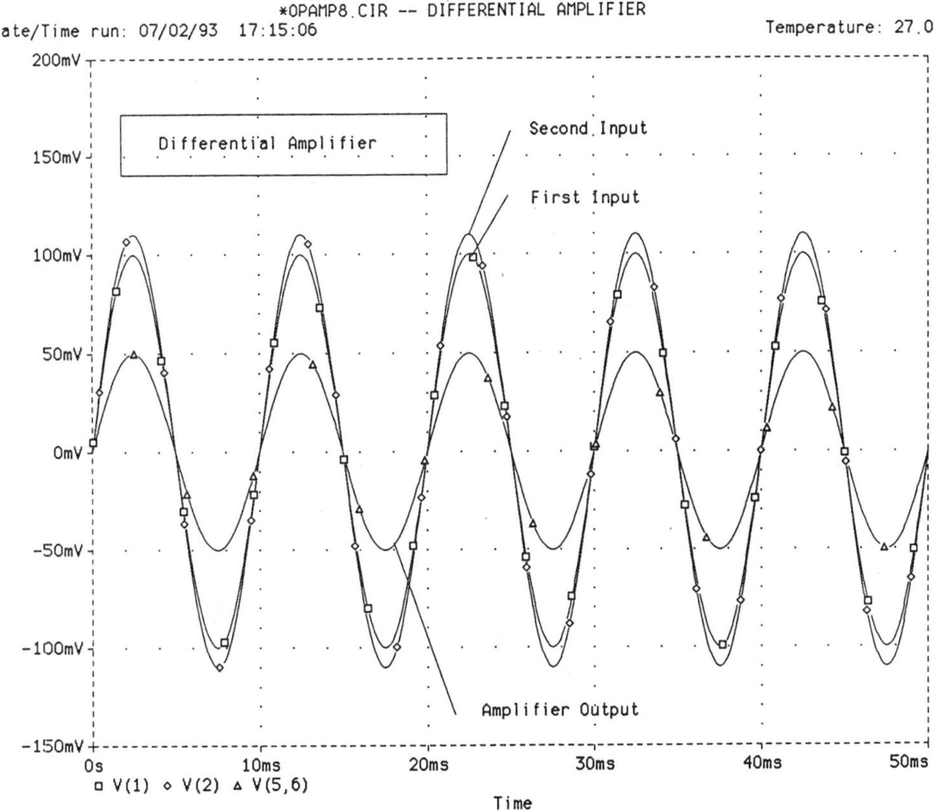

Figure 5.14

In Figure 5.14, after the three traces for the two input waveforms and the output wave were plotted, labels were added. In fact, the labels are not necessary because the traces are identified by node numbers at the bottom of the plot. This is adequate but not very convenient. Labels allow the reader to identify traces quickly without noting identifying squares, diamonds, and triangles or referring to the circuit description for node numbers. Notes, qualifications, and titles can also be included on the graph.

To add the labels to the graph, the "Label" option was intuitively chosen from the main menu. Choosing "Text" from the submenu produces an invitation to type in the text to be included on the graph. After the text is typed and the "Enter" key pressed, the text appears in the center of the graph. The text can then be moved to a more convenient location using a mouse and a standard click and drag operation. The "Line" option works as expected by pressing the left mouse button when the arrow points to the beginning of the line, dragging to the end of the line, and releasing the button. The "Box," "Circle," and "Ellipse" operations are equally intuitive.

6

Bipolar Junction Transistor Circuits

Like diodes and op amps, bipolar junction transistors (BJTs) vary widely in a long list of characteristics. To simulate circuits containing BJTs successfully, PSpice must have access to values for p-n saturation current, junction capacitances, forward maximum beta, and almost 40 other parameters; besides knowing whether the transistor is constructed as a PNP or NPN.

The evaluation library contains this information for four bipolar junction transistors. Before the user can begin to analyze circuits containing BJTs, it is necessary to create another library containing these four transistor models in order to use the .INC statement rather than .LIB. As before, use the DOS "copy" command to make a copy of EVAL.LIB called TRAN.LIB. Then delete all the contents except the following four transistors.

```
.model Q2N2222A NPN(Is=14.34f Xti=3 Eg=1.11 Vaf=74.03 Bf=255.9 Ne=1.307
+      Ise=14.34f Ikf=.2847 Xtb=1.5 Br=6.092 Nc=2 Isc=0 Ikr=0 Rc=1
+      Cjc=7.306p Mjc=.3416 Vjc=.75 Fc=.5 Cje=22.01p Mje=.377 Vje=.75
+      Tr=46.91n Tf=411.1p Itf=.6 Vtf=1.7 Xtf=3 Rb=10)
*     National      pid=19         case=TO18
*     88-09-07 bam       creation

.model Q2N2907A PNP(Is=650.6E-18 Xti=3 Eg=1.11 Vaf=115.7 Bf=231.7
Ne=1.829
+      Ise=54.81f Ikf=1.079 Xtb=1.5 Br=3.563 Nc=2 Isc=0 Ikr=0 Rc=.715
+      Cjc=14.76p Mjc=.5383 Vjc=.75 Fc=.5 Cje=19.82p Mje=.3357 Vje=.75
+      Tr=111.3n Tf=603.7p Itf=.65 Vtf=5 Xtf=1.7 Rb=10)
*     National      pid=63         case=TO18
*     88-09-09 bam       creation

.model Q2N3904 NPN(Is=6.734f Xti=3 Eg=1.11 Vaf=74.03 Bf=416.4 Ne=1.259
+      Ise=6.734f Ikf=66.78m Xtb=1.5 Br=.7371 Nc=2 Isc=0 Ikr=0 Rc=1
+      Cjc=3.638p Mjc=.3085 Vjc=.75 Fc=.5 Cje=4.493p Mje=.2593 Vje=.75
+      Tr=239.5n Tf=301.2p Itf=.4 Vtf=4 Xtf=2 Rb=10)
```

```
*     National     pid=23        case=TO92
*     88-09-08 bam     creation

.model Q2N3906 PNP(Is=1.41f Xti=3 Eg=1.11 Vaf=18.7 Bf=180.7 Ne=1.5 Ise=0
+     Ikf=80m Xtb=1.5 Br=4.977 Nc=2 Isc=0 Ikr=0 Rc=2.5 Cjc=9.728p
+     Mjc=.5776 Vjc=.75 Fc=.5 Cje=8.063p Mje=.3677 Vje=.75 Tr=33.42n
+     Tf=179.3p Itf=.4 Vtf=4 Xtf=6 Rb=10)
*     National     pid=66        case=TO92
*     88-09-09 bam     creation
```

6.1 Transistor DC Load Lines

Most textbooks begin the discussion of transistors by explaining the orientation of the doped regions and the physical laws that govern the flow of current through these regions. To make the transistor perform a useful function, it is necessary to forward bias the base-emitter junction and reverse bias the base-collector junction.

Any of several configurations of resistors can be used to achieve the proper biasing. In each configuration, the goal is to provide a voltage at the base that results in a collector-emitter voltage near the center of the DC load line. In practice, some configurations are more successful at creating a stable biasing point; and for this reason, texts usually focus on voltage-divider biasing more than on the other methods. Example 6.1 illustrates voltage-divider biasing.

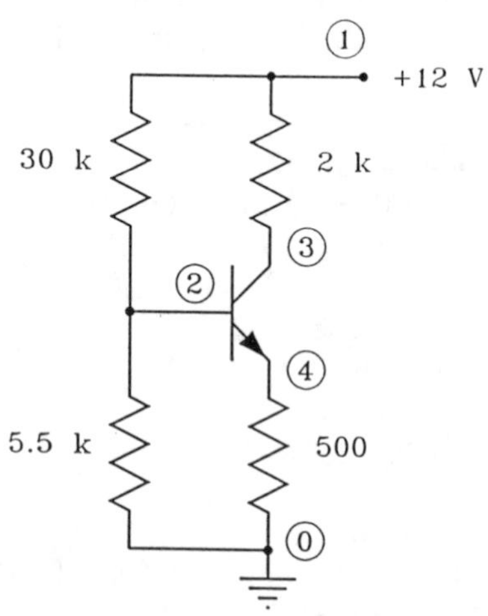

Example 6.1

To bias the transistor properly using the voltage-divider configuration, a DC load line is plotted on a graph. The current through the transistor is plotted as a function of the voltage across the transistor. The DC load line is drawn between two extreme points of the transistor response. The first point (saturation) represents the "on" state of the transistor when it is virtually transparent, passing maximum current and dropping very little voltage. The opposite end of the line (cutoff) represents the "off" condition, when the transistor passes minimal current and drops nearly the entire source voltage.

Since the "on" transistor always drops a little voltage and the "off" transistor passes a little current, the end points of the load line are open to interpretation. There are two standard ways to determine the DC load line end points, resulting in two different lines. This leads to some confusion that PSpice can be used to resolve.

The two load lines result from two different methods for driving the transistor from cutoff to saturation. These methods are illustrated in Examples 6.1A and 6.1B along with the circuit descriptions for PSpice. In Example 6.1A, the swing from cutoff to saturation is achieved by varying the resistor R2. In Example 6.1B, the emitter resistor is varied.

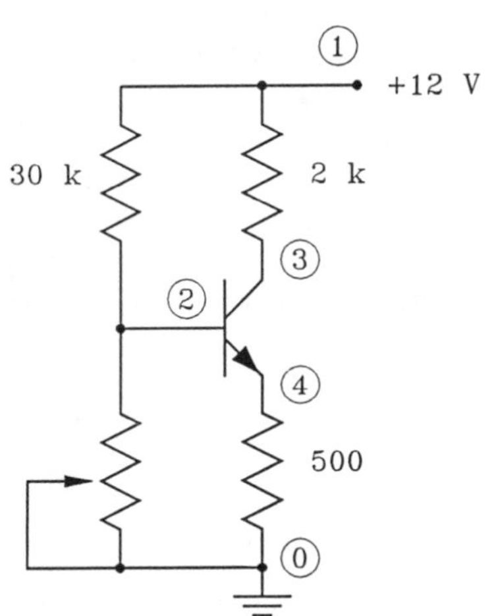

Example 6.1A

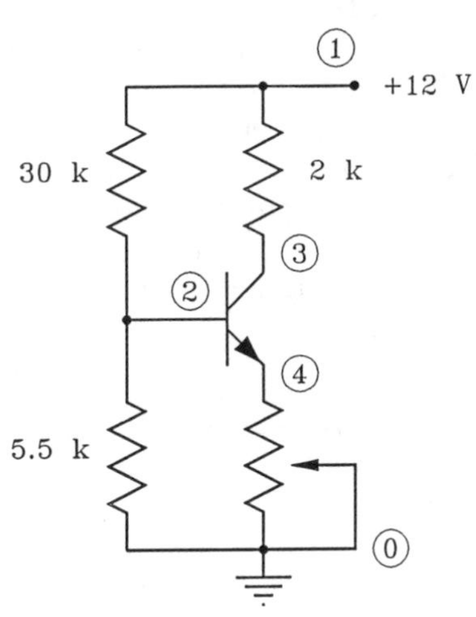

Example 6.1B

```
*TRAN1A.CIR–
*DC BIAS, VARIABLE R2
.INC \SPICE\TRAN.LIB
V1 1 0 12
R1 1 2 30K
RCOL 1 3 2K
REMIT 4 0 500
.PARAM VALUE = 1K
.STEP PARAM(VALUE) 1K 12K 500
R2 2 0 {VALUE}
Q1 3 2 4 Q2N2222A
.DC V1 12 12 1
.PRINT DC I(RCOL) V(3,4)
.END
```

```
*TRAN1B.CIR–
*DC BIAS, VARIABLE REMIT
.INC \SPICE\TRAN.LIB
V1 1 0 12
R1 1 2 30K
RCOL 1 3 2K
REMIT 4 0 {VALUE}
.PARAM VALUE = 1K
.STEP PARAM(VALUE) 200 4K 200
R2 2 0 5.5K
Q1 3 2 4 Q2N2222A
.DC V1 12 12 1
.PRINT DC I(RCOL) V(3,4)
.END
```

The first item to note in the circuits is the description of the transistor. Bipolar transistor names always begin with a "Q." The general form is

Q<name> <collector> <base> <emitter> <model>

As in the case of subcircuits, the order of the transistor nodes is critically important.

For both circuits, the analysis is a DC sweep for one voltage, beginning and ending at 12 Volts. .PARAM creates the variable "value" and .STEP defines the range for the variable and the increment. The requested output is the collector current and the voltage across the transistor. The circuits are in all respects identical except for the location and the range of the variable resistor. In TRAN1A.CIR, the variable resistor is the R2 biasing resistor, and the emitter resistor is fixed at 500 ohms. In TRAN1B.CIR, R2 is fixed at 5.5 K and the emitter resistor varies. In each circuit, the range for the variable resistor has been kept sufficiently wide to drive the transistor from cutoff to saturation. This is exactly the procedure that would be carried out in the lab to determine the DC load line for each circuit.

The output information is printed in each circuit's .OUT file as collector current and collector-emitter voltage for each value of the variable resistor. Since a separate analysis is carried out for each resistor value of the variable resistor, the output file requires some editing. After a little rearrangement the two output files can yield the following tables.

*TRAN1A.CIR – DC BIAS, VARIABLE R2

V1	I(RCOL)	V(3,4)		
1.200E+01	5.240E−08	1.200E+01	PARAM VALUE =	1.0000E+03
1.200E+01	3.338E−05	1.192E+01	PARAM VALUE =	1.5000E+03
1.200E+01	2.735E−04	1.132E+01	PARAM VALUE =	2.0000E+03
1.200E+01	5.696E−04	1.057E+01	PARAM VALUE =	2.5000E+03
1.200E+01	8.687E−04	9.825E+00	PARAM VALUE =	3.0000E+03
1.200E+01	1.162E−03	9.091E+00	PARAM VALUE =	3.5000E+03
1.200E+01	1.448E−03	8.376E+00	PARAM VALUE =	4.0000E+03
1.200E+01	1.725E−03	7.683E+00	PARAM VALUE =	4.5000E+03
1.200E+01	1.993E−03	7.013E+00	PARAM VALUE =	5.0000E+03
1.200E+01	2.252E−03	6.364E+00	PARAM VALUE =	5.5000E+03
1.200E+01	2.502E−03	5.738E+00	PARAM VALUE =	6.0000E+03
1.200E+01	2.744E−03	5.132E+00	PARAM VALUE =	6.5000E+03
1.200E+01	2.978E−03	4.546E+00	PARAM VALUE =	7.0000E+03
1.200E+01	3.204E−03	3.980E+00	PARAM VALUE =	7.5000E+03
1.200E+01	3.423E−03	3.432E+00	PARAM VALUE =	8.0000E+03
1.200E+01	3.635E−03	2.902E+00	PARAM VALUE =	8.5000E+03
1.200E+01	3.840E−03	2.389E+00	PARAM VALUE =	9.0000E+03
1.200E+01	4.038E−03	1.893E+00	PARAM VALUE =	9.5000E+03
1.200E+01	4.230E−03	1.413E+00	PARAM VALUE =	10.0000E+03
1.200E+01	4.416E−03	9.471E−01	PARAM VALUE =	10.5000E+03

```
1.200E+01   4.596E-03   4.960E-01      PARAM VALUE =   11.0000E+03
1.200E+01   4.732E-03   1.551E-01      PARAM VALUE =   11.5000E+03
1.200E+01   4.746E-03   1.146E-01      PARAM VALUE =   12.0000E+03

*TRAN1B.CIR-DC BIAS, VARIABLE REMIT

    V1          I(RCOL)       V(3,4)

1.200E+01   5.116E-03   7.387E-01      PARAM VALUE = 200
1.200E+01   2.766E-03   5.355E+00      PARAM VALUE = 400
1.200E+01   1.899E-03   7.055E+00      PARAM VALUE = 600
1.200E+01   1.448E-03   7.938E+00      PARAM VALUE = 800
1.200E+01   1.171E-03   8.479E+00      PARAM VALUE =    1.0000E+03
1.200E+01   9.841E-04   8.844E+00      PARAM VALUE =    1.2000E+03
1.200E+01   8.488E-04   9.107E+00      PARAM VALUE =    1.4000E+03
1.200E+01   7.464E-04   9.305E+00      PARAM VALUE =    1.6000E+03
1.200E+01   6.663E-04   9.460E+00      PARAM VALUE =    1.8000E+03
1.200E+01   6.018E-04   9.585E+00      PARAM VALUE =    2.0000E+03
1.200E+01   5.488E-04   9.687E+00      PARAM VALUE =    2.2000E+03
1.200E+01   5.045E-04   9.772E+00      PARAM VALUE =    2.4000E+03
1.200E+01   4.668E-04   9.845E+00      PARAM VALUE =    2.6000E+03
1.200E+01   4.345E-04   9.907E+00      PARAM VALUE =    2.8000E+03
1.200E+01   4.063E-04   9.960E+00      PARAM VALUE =    3.0000E+03
1.200E+01   3.816E-04   1.001E+01      PARAM VALUE =    3.2000E+03
1.200E+01   3.598E-04   1.005E+01      PARAM VALUE =    3.4000E+03
1.200E+01   3.403E-04   1.009E+01      PARAM VALUE =    3.6000E+03
1.200E+01   3.229E-04   1.012E+01      PARAM VALUE =    3.8000E+03
1.200E+01   3.072E-04   1.015E+01      PARAM VALUE =    4.0000E+03
```

These data would be more useful if they were displayed graphically. Although PSpice is able to graph one output variable as a function of another, it cannot graph values from different analyses. To illustrate the DC load line that results from these voltages and currents, the data must be graphed either by hand or by computer using graphic software. Fortunately, the data can be given to virtually any spread sheet and the collector-emitter voltage used as the independent variable. If a spread sheet is not available, writing a program to examine a data file and graph the coordinate data is not difficult, as shown in Figure 6.1.

The data from TRAN1A.CIR indicate that the load line has a cutoff voltage of 12 Volts or Vcc. The saturation current is 4.8 milliamps or Vcc/(Rcol + Remit). The trace for TRAN1B.CIR has a cutoff voltage of approximately 10.7 Volts—the source voltage minus the voltage dropped across the emitter resistor. The saturation current is approximately 5.35 milliamps or 10.7V/Rcol.

The logical question is which line is the "real" DC load line. The answer is that one is just as real as the other. The important point is to notice that they intersect at very near the mid-point of each line. Since this center point is where we prefer to bias small signal amplifiers, the two lines lead to the same biasing.

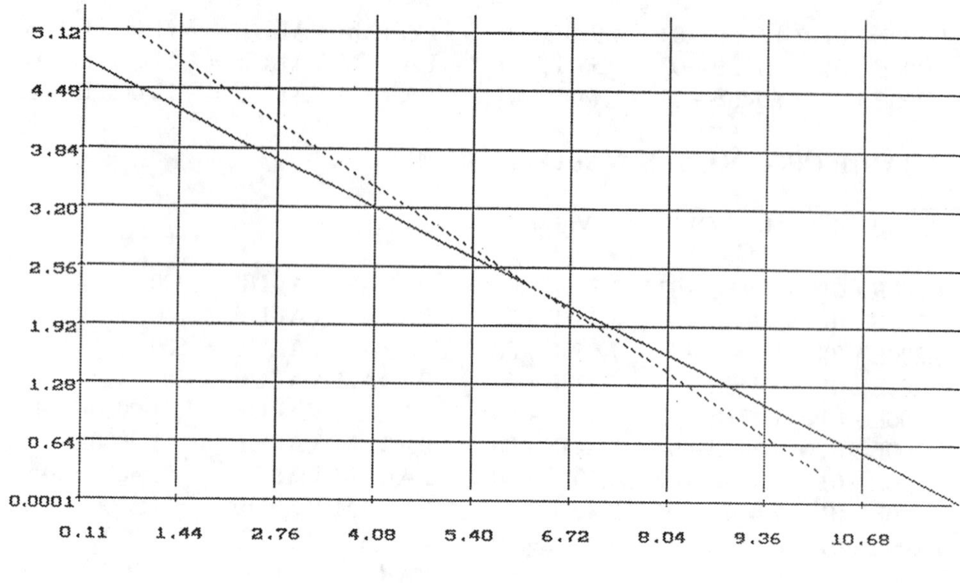

Figure 6.1

6.2 DC Load Line in Collector Feedback Biasing

The procedure for finding a DC load line by varying a resistor to drive a transistor from cutoff to saturation applies to other biasing configurations. In Example 6.3 the feedback resistor is varied from 100 kohms to 1 Megohm to determine the DC load line and the resistor value that will produce a drop of approximately one-half the supply voltage across the transistor.

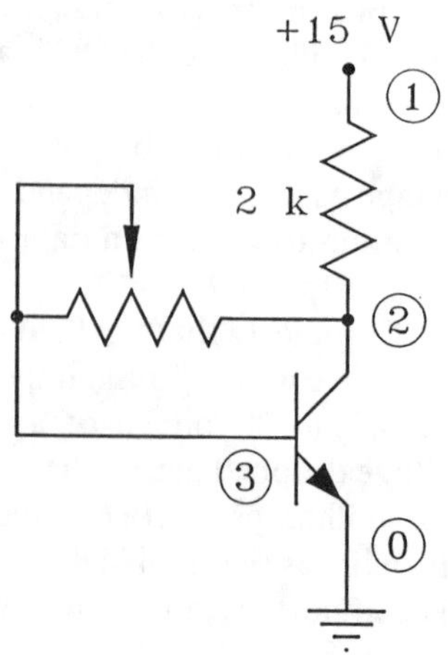

Example 6.2

```
*TRAN2.CIR—COLLECTOR-FEEDBACK BIASING
.INC \SPICE\TRAN.LIB
V1 1 0 15
RCOL 1 2 2K
Q1 2 3 0 Q2N2222A
.PARAM VALUE = 1K
.STEP PARAM(VALUE) 100K 1MEG 50K
RFEEDBACK 2 3 {VALUE}
.DC V1 15 15 1
.PRINT DC I(RCOL) V(2)
.END
```

PSpice analysis provides the following values. Again, generous editing is necessary to assemble the data into table form.

V1	I(RCOL)	V(2)	
1.500E+01	5.579E−03	3.842E+00	PARAM VALUE = 100.0000E+03
1.500E+01	5.037E−03	4.927E+00	PARAM VALUE = 150.0000E+03
1.500E+01	4.592E−03	5.815E+00	PARAM VALUE = 200.0000E+03
1.500E+01	4.221E−03	6.559E+00	PARAM VALUE = 250.0000E+03
1.500E+01	3.905E−03	7.191E+00	PARAM VALUE = 300.0000E+03
1.500E+01	3.632E−03	7.736E+00	PARAM VALUE = 350.0000E+03
1.500E+01	3.394E−03	8.211E+00	PARAM VALUE = 400.0000E+03
1.500E+01	3.185E−03	8.629E+00	PARAM VALUE = 450.0000E+03
1.500E+01	3.000E−03	9.000E+00	PARAM VALUE = 500.0000E+03
1.500E+01	2.834E−03	9.332E+00	PARAM VALUE = 550.0000E+03
1.500E+01	2.685E−03	9.630E+00	PARAM VALUE = 600.0000E+03
1.500E+01	2.551E−03	9.899E+00	PARAM VALUE = 650.0000E+03
1.500E+01	2.428E−03	1.014E+01	PARAM VALUE = 700.0000E+03
1.500E+01	2.317E−03	1.037E+01	PARAM VALUE = 750.0000E+03
1.500E+01	2.215E−03	1.057E+01	PARAM VALUE = 800.0000E+03
1.500E+01	2.121E−03	1.076E+01	PARAM VALUE = 850.0000E+03
1.500E+01	2.034E−03	1.093E+01	PARAM VALUE = 900.0000E+03
1.500E+01	1.954E−03	1.109E+01	PARAM VALUE = 950.0000E+03
1.500E+01	1.880E−03	1.124E+01	PARAM VALUE = 1.0000E+06

From an examination of the table it is apparent that a resistor value between 300 kohms and 350 kohms will provide the required 7.5 Volts across the transistor.

6.3 Small-Signal Amplifiers

Of course, properly biasing a transistor is only the beginning. Once it is biased, we would like to have it do something useful—amplify a signal, for example. Amplifiers are usually divided into small-signal and power amplifiers, although there is no firm definition of either. Some authors loosely define a small-signal amplifier as one that has a peak-to-peak output voltage of less than 10% of the source voltage. In practice, a small-signal amplifier can be thought of as an amplifier that uses such a small portion of the transistor transconductance curve that harmonic distortions are negligible. By simulating a spectrum analyzer, PSpice can display graphically any output harmonics that exist.

Example 6.3 uses a transistor biased as in Example 6.1 to amplify a 1 millivolt, 10 kHz input signal. The complete circuit includes a signal generator, a load resistor, and coupling and bypass capacitors. The capacitors' values have been chosen so the capacitors are virtually transparent at 10 kHz.

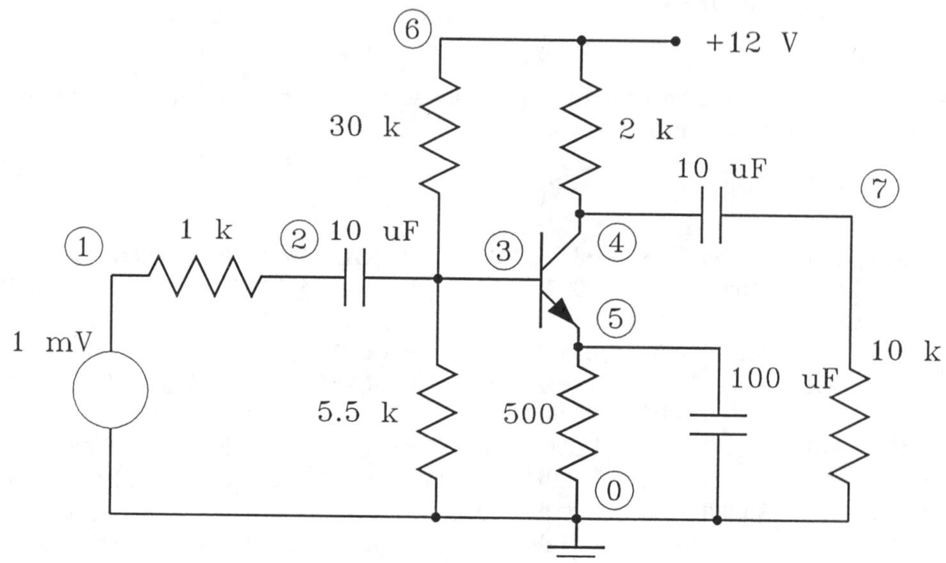

Example 6.3

Example 6.3 has been analyzed four times. The first analysis is a time-based analysis for the circuit as initially described. In the second analysis, an AC source and AC analysis are substituted for the SIN function and the transient analysis to provide a frequency-based analysis at the single frequency of 10 kHz. This provides AC voltage and current information that can be compared to VOM readings. For the third analysis, a frequency sweep is used to determine the frequency response of the amplifier. Finally, a DC analysis is performed to check biasing voltages. For each analysis, only the lines between the "******" markers need to be changed.

```
*TRAN3.CIR—SMALL SIGNAL AMPLIFIER
.INC \SPICE\TRAN.LIB
VSUPPLY 6 0 12
CIN 2 3 10U
COUT 4 7 10U
CBYPASS 5 0 100U
RIN 1 2 1K
```

```
R1 6 3 30K
R2 3 0 5.5K
RCOL 6 4 2K
REMIT 5 0 500
RLOAD 7 0 10K
******
Q1 4 3 5 Q2N2222A
VIN 1 0 SIN ( 0 1M 10K )
.TRAN 10U .5M 0 .3U
.PROBE
******
.END
```

Time-Based Analysis Since the magnitude of the AC waveforms in the circuits is small compared to that of the DC biasing voltages, placing several traces on the same plot would result in widely separated sinusoids that are nearly flat lines. Consequently, three separate plots are used in Figure 6.2 to display the base voltage, the collector voltage, and the load voltage, respectively, from top to bottom. Besides providing amplitude information, the plots illustrate the phase inversion between the base and collector voltages.

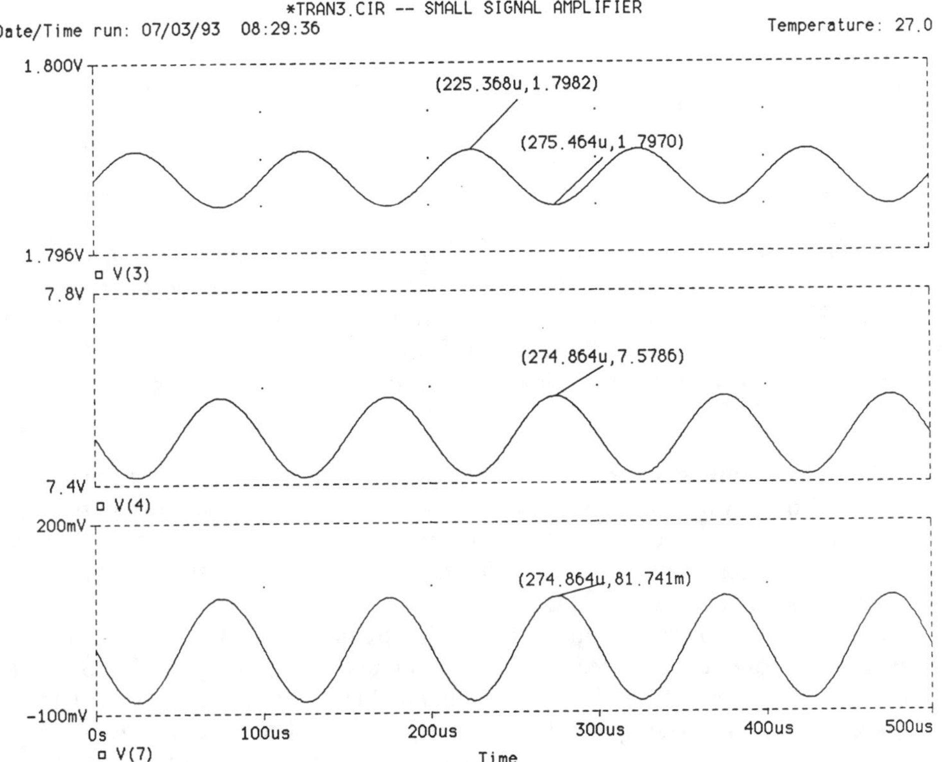

Figure 6.2

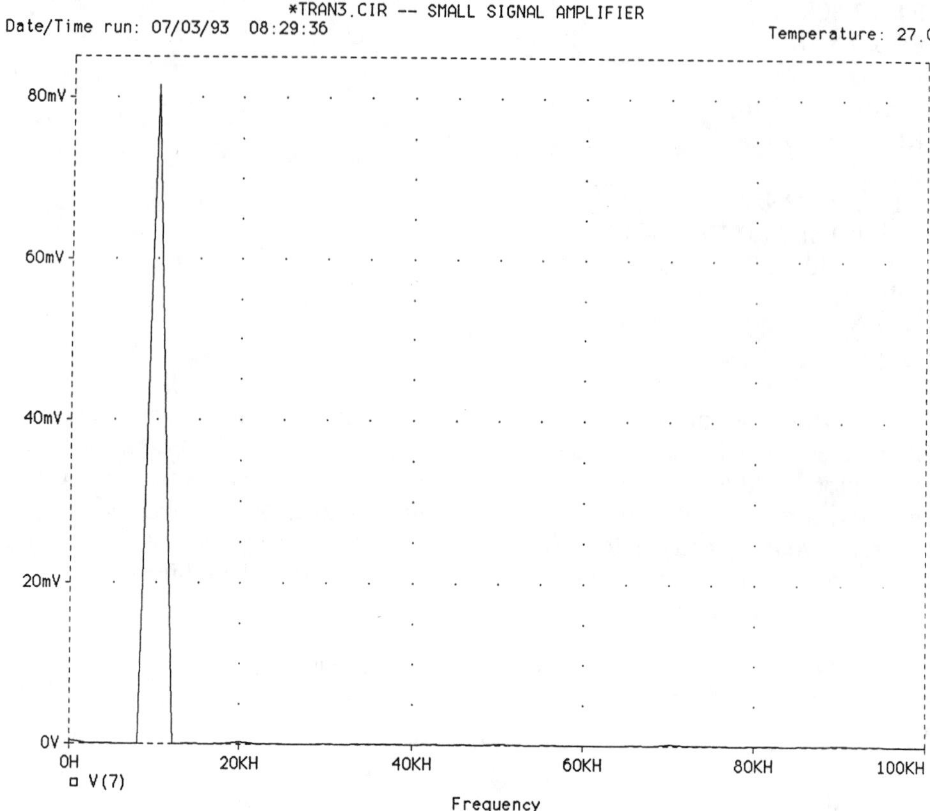

Figure 6.3

Ideally, a small-signal amplifier should not add any harmonic distortion to the output signal. Spectral information can be obtained by graphing only the load voltage and choosing the "X-axis" option from the main menu. Selecting "Fourier" from the submenu results in the plot shown in Figure 6.3. While the second and seventh harmonics are present, their amplitudes are negligible and barely exceed background noise.

Single-Frequency Analysis Using the .AC statement and requesting analysis at a single frequency instructs PSpice to provide AC peak voltage and current readings. This can be more convenient than estimating peak values from a graph and hoping all transient effects have settled out. The voltage values of interest are at the base, at the collector, and across the load.

The AC beta value can be calculated for the circuit if values for base and collector current are requested. This requires a small amount of ingenuity. While PSpice will provide current through a resistor, it will not provide current through a three-terminal device—i.e., I(RLOAD) is acceptable but I(Q1) is not. The trick is to insert a zero-valued two-terminal device into the branch where the current is needed. In this amplifier (Example 6.4), a voltage source of zero volts, VFICTION1, is placed between the intersection of R1 and R2, and the base of the transistor. The intersection of the biasing resistors is called node 3, and the base of the transistor is renamed node 33. Since the voltage source VFICTION1 provides no voltage, it has no effect on the circuit; but it does provide a two-terminal device through which current can be measured. Similarly, VFICTION2 should be placed between the intersection of RCOL and COUT, and the collector. The intersection remains node 4, and the collector becomes node 44.

It is worth noting that while the .PRINT TRAN and .PRINT AC statements will not accept a request for current through a three-terminal device, PROBE will accept such requests. In PROBE, typing IB(Q1) after "Add trace" is selected will create a plot of the current into the base of the transistor, as shown in Example 6.4. Collector and emitter current are indicated by IC(Q1) and IE(Q1), respectively.

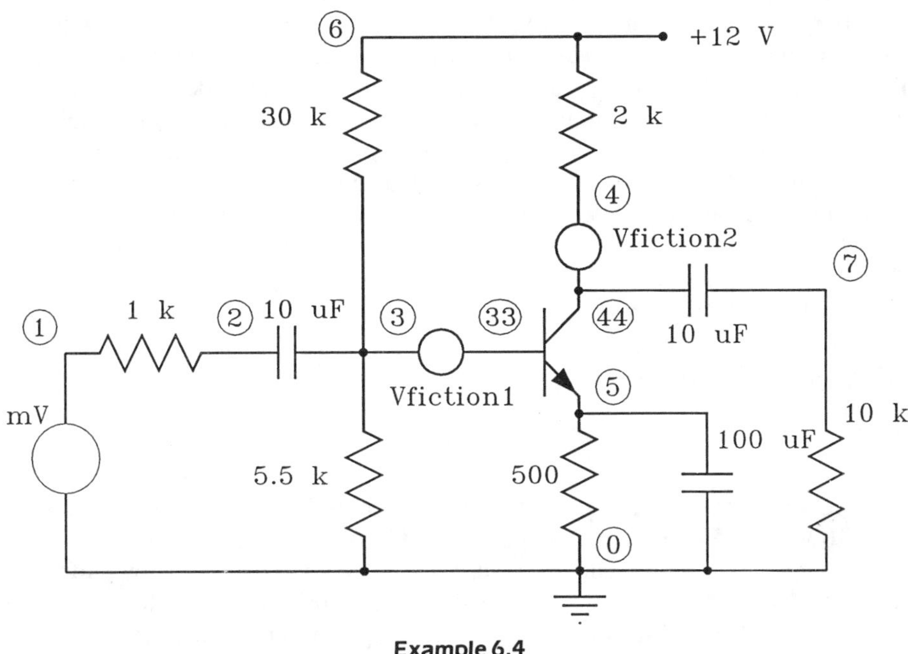

Example 6.4

To include the fictional voltage sources and the proper analysis statements, the circuit description lines between the markers must be changed to

```
******
Q1 44 33 5 Q2N2222A
VFICTION1 3 33 0
VFICTION2 4 44 0
VIN 1 0 AC 1M
.AC LIN 1 10K 10K
.PRINT AC V(3) V(4) V(7) I(VFICTION1) I(VFICTION2)
******
```

The .OUT file contains the requested voltages and currents.

FREQ	V(33)	V(44)	V(7)	I(VFICTION1)	I(VFICTION2)
1.000E+04	5.954E−04	8.149E−02	8.149E−02	2.773E−07	4.889E−05

The voltage at the base indicates that approximately .4 millivolts of the signal was lost to the resistance of the signal generator. Since the collector voltage and the load voltage are identical, the output coupling capacitor must be transparent. Dividing the collector current by the base current yields an AC beta of 176. This is, of course, a typical value used for analysis purposes. Real beta values vary widely even among "identical" transistors.

Frequency Sweep Analysis Transistor amplifiers do not perform equally well at all frequencies. Coupling and bypass capacitors cause frequencies below the cutoff frequency to be suppressed, and internal capacitances in the transistor cause a high frequency rolloff. By sweeping the generator through a wide range of frequencies, the analysis software can provide PROBE with the data needed to graph a frequency response curve. For the frequency sweep, the fictional voltage sources are omitted, and the transistor is given its original node designations. The following lines should be included between the markers in the circuit description.

```
******
Q1 4 3 5 Q2N2222A
VIN 1 0 AC 1M
.AC DEC 100 10 10MEG
.PROBE
******
```

The .AC statement uses a logarithmic distribution of 100 points over a range of 10 Hz to 10 MHz. Notice that in the circuit description megahertz is indicated by "MEG."

The frequency response curve shown in Figure 6.4 was developed by plotting the output voltage and using the "Cursor" submenu to label the points. Returning to the main menu and choosing "Label" provides the opportunity to insert "AMPLIFIER BANDWIDTH" and the arrows.

DC Analysis Analyzing the circuit one last time provides numerical values for the DC biasing levels at critical points in the circuit. For trouble shooting purposes, a technician would be interested in the DC levels at the base, emitter, and collector. The DC beta might also be of some interest, so PSpice has been instructed to display DC currents into the base and collector. As in the second analysis, the fictional voltage supplies must be inserted just before the base and the collector in order to find the current entering the three-terminal device.

To perform the required DC analysis, the following must be inserted into the circuit description.

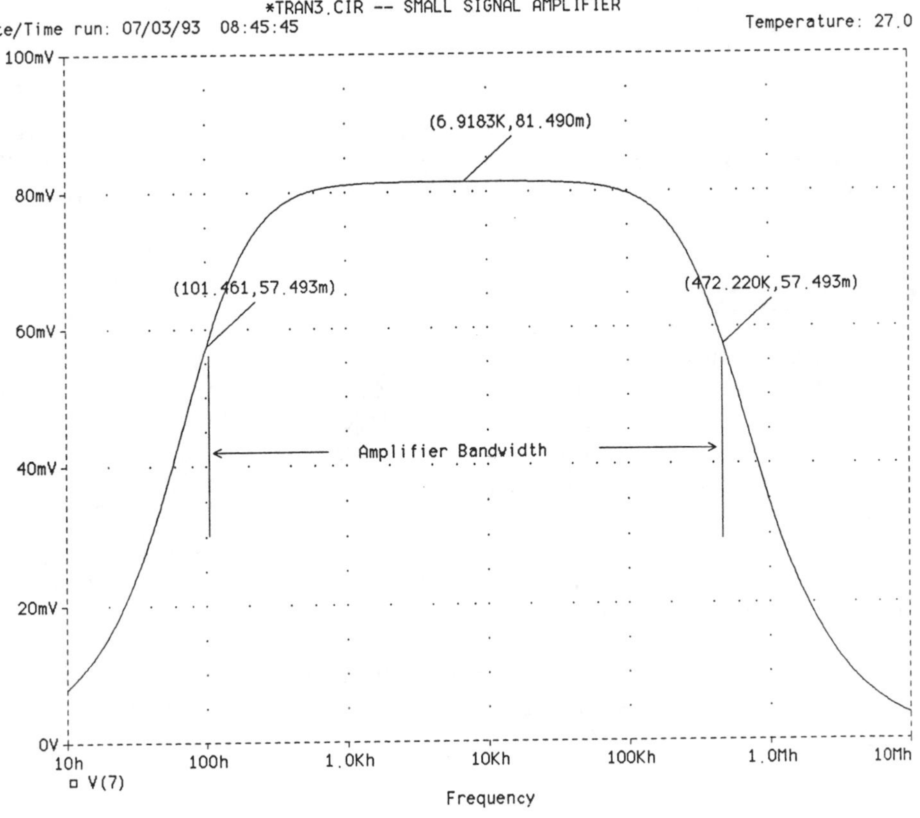

Figure 6.4

```
******
Q1 44 33 5 Q2N2222A
VFICTION1 3 33 0
VFICTION2 4 44 0
VIN 1 0 AC 1M
.DC VSUPPLY 12 12 1
.PRINT DC V(33) V(44) V(7) V(5)
.PRINT DC I(VFICTION1) I(VFICTION2)
******
```

After deleting all but the relevant information from the .OUT file, we are left with the following data.

VSUPPLY	V(33)	V(44)	V(7)	V(5)
1.200E+01	1.798E+00	7.497E+00	0.000E+00	1.132E+00

VSUPPLY	I(VFICTION1)	I(VFICTION2)
1.200E+01	1.325E−05	2.252E−03

As expected, the DC load voltage is zero, and there is a voltage drop of .666 Volts across the base-emitter diode. The voltage from collector to emitter is 7.497 − 1.132 = 6.365, or about the center of the DC load line. Dividing the collector current by the base current gives a DC beta of 170.

6.4 Transistor Modeling

Like all electronic components, transistors can be modeled in many ways depending on the level of accuracy required from the model and the amount of complexity the writer of the model is willing or able to include. Even a simple resistor can be modeled with its associated end-to-end capacitance and lead inductance or without it. The transistor model used by PSpice is extremely complex, to allow for accurate simulation in a wide variety of circuit applications.

In electronics textbooks, students usually encounter simpler models included to make transistor behavior clear. The circuit described here, based on Example 6.3, models only the AC behavior of the transistor in the bandpass region.

A transistor is a dependent current device. This means that the amount of current through the transistor, flowing from the collector to the emitter, is dependent upon or controlled by some other current or voltage. In Example 6.5, the transistor current is dependent upon the voltage generated across the base-emitter junction.

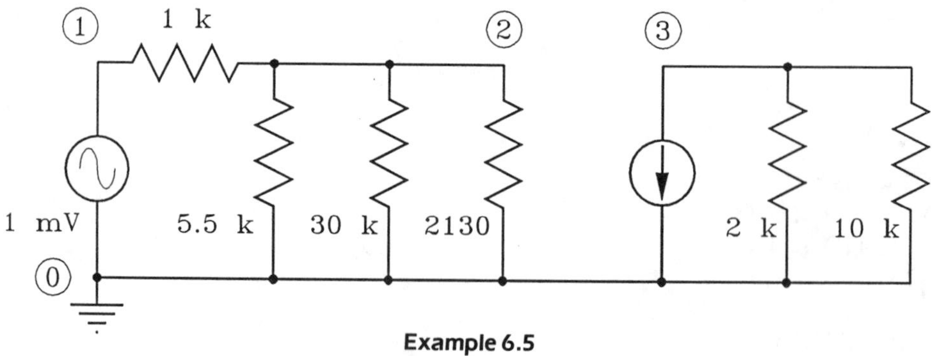

Example 6.5

The signal generator is the independent source. Values for the generator resistance and the biasing resistors are taken from Example 6.3. The unfamiliar resistance is the 2130 ohm resistor. This value is the AC emitter resistance reflected back to the input circuit. The value is equal to beta multiplied by the base-emitter AC resistance. Beta was previously determined to be 176. The base-emitter AC resistance (r_e') can be determined roughly by dividing 25 mV by the DC emitter current or more accurately by dividing the base-emitter voltage by the base-emitter current—i.e., by

Ohm's Law. These values can be found by performing a PSpice single-frequency analysis on Example 6.3 and asking for the voltage from node 33 to node 5. The necessary current is the sum of I(VFICTION1) and I(VFICTION2).

FREQ	V(33,5)	I(VFICTION1)	I(VFICTION2)
1.000E+04	5.953E−04	2.773E−07	4.889E−05

The resulting value for r_e' is 12.1 ohms. Multiplying this resistance by beta gives 2130 ohms.

PSpice supports all four forms of dependent sources. Each is designated by a letter from E to H.

E – Voltage-controlled voltage source
F – Current-controlled current source
G – Voltage-controlled current source
H – Current-controlled voltage source

Example 6.5 uses a voltage-controlled current source. The general form is

G<name> <+ node> <− node> <+ controlling node> <− controlling node>
\+ ***<transconductance>***

<+ node> and <− node> refer to the dependent current source. Positive current is considered to flow from the + node to the − node **through the source.** The controlling nodes mark the voltage that controls the current source. The transconductance is the inverse of the base-emitter AC resistance, in this case, .0826. The circuit description for Example 6.5 is

```
*TRAN4.CIR—AC MODEL
VIN 1 0 AC 1M
RSOURCE 1 2 1K
R2 2 0 5.5K
R1 2 0 30K
REMIT 2 0 2130
GTRAN 3 0 2 0 .0826
RCOL 3 0 2K
RLOAD 3 0 10K
.AC LIN 1 10K 10K
.PRINT AC V(2) V(3)
.END
```

The output file for this somewhat crude model contains a load voltage value that compares well with the PSpice simulation using the transistor library model.

FREQ	V(2)	V(3)
1.000E+04	5.936E−04	8.172E−02

Example 6.6 uses the same model except for the use of a current-controlled current device. Although the dependent current source should be controlled by the current through the 2130 ohm resistor, PSpice requires that current-controlled current sources be controlled by the current through a voltage device. To accommodate PSpice, a zero-valued voltage device has been inserted in series with the 2130 ohm resistor.

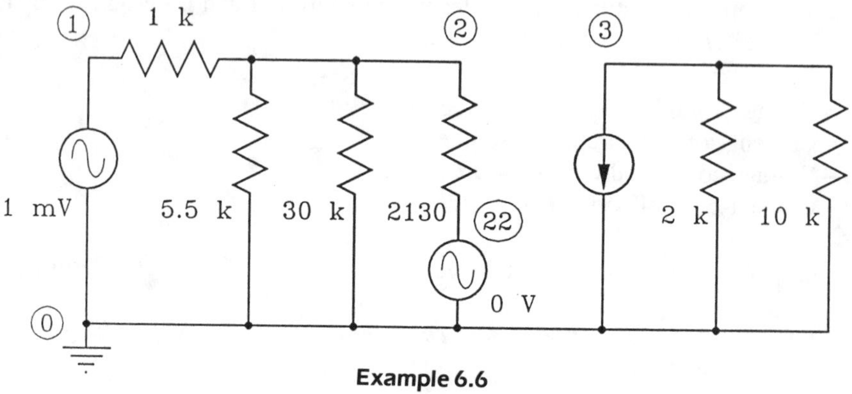

Example 6.6

The general form for a current-controlled current device is

F<name> <+ node> <− node> < controlling voltage device> <current gain>

The controlling voltage device, VBASE, is positioned between nodes 22 and ground. The amplifier current gain as calculated from the results of Example 6.3 is 176. The resulting circuit description and analysis output are given below.

```
*TRAN4.CIR−AC MODEL
VIN 1 0 AC 1M
RSOURCE 1 2 1K
R2 2 0 5.5K
R1 2 0 30K
REMIT 2 22 2130
VBASE 22 0 AC 0
FTRAN 3 0 VBASE 176
RCOL 3 0 2K
RLOAD 3 0 10K
.AC LIN 1 10K 10K
.PRINT AC V(2) V(3)
.END
```

FREQ	V(2)	V(3)
1.000E+04	5.936E−04	8.175E−02

Again, the load voltage compares well with the results from the other models.

The general forms and complete descriptions of all four dependent sources are given in Appendix C.

6.5 Cascaded Amplifiers

Real devices seldom consist of a single transistor circuit. More commonly, several transistors are cascaded to create a system that performs a required function. To be useful, PSpice must be able to correctly analyze multiple transistor systems. Example 6.7, while not a very practical amplifier, illustrates the use of PSpice to analyze a system of three different amplifier stages. In order to maintain small-signal operation, the first two stages are heavily swamped. Stage two uses the 2N2907A PNP transistor in the upside down, positive-source configuration. Stage three is a common collector amplifier.

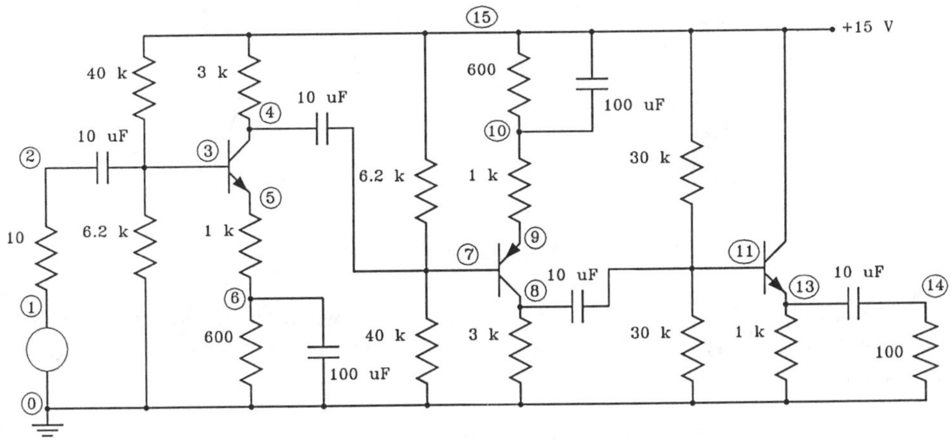

Example 6.7

No new PSpice features are introduced in this section, but the circuit deserves a few comments. Nodes for the PNP transistor are given in the same collector-base-emitter order used for NPN transistors. The input for the PNP (node 7) has been biased to about 13 Volts, compared to 2 Volts for the NPN. The graphic output indicates that the second common emitter amplifier stage has brought the output back into phase with the input. For clarity, components in the first stage have been given a "−1" suffix, stage-two components have a "−2" suffix, and so on.

The circuit description has been written to provide AC voltage levels and a time-based graphic analysis, as shown in Figure 6.5. DC and frequency response information could have been requested using the techniques described in section 6.3 for a

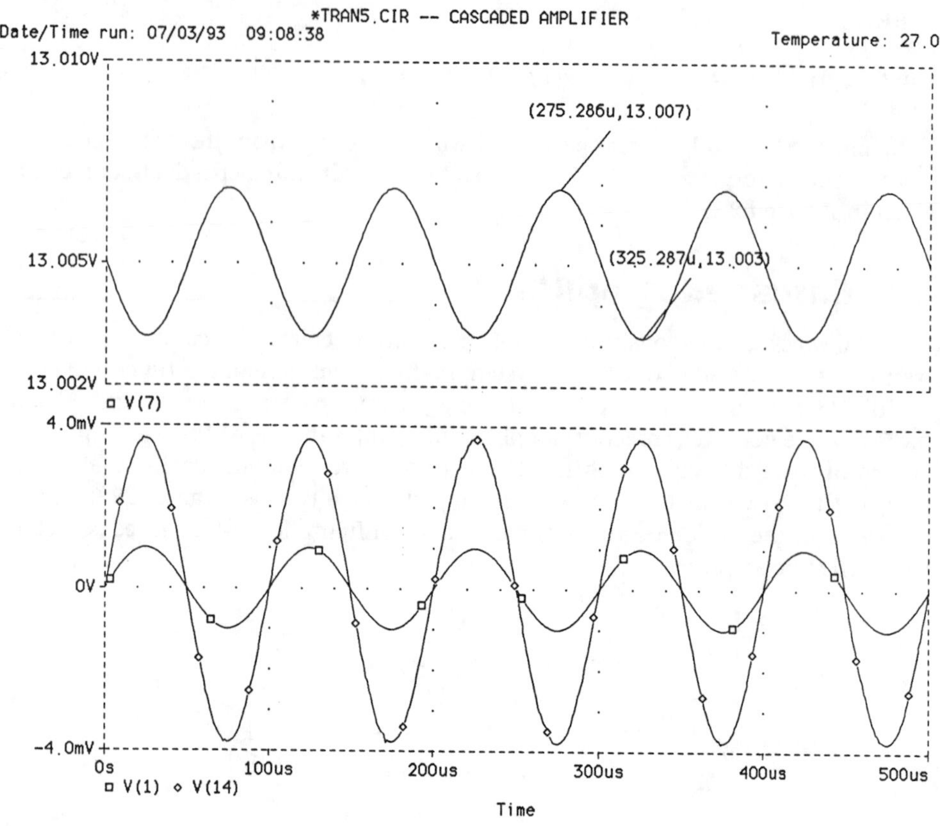

Figure 6.5

single-stage amplifier. The circuit description is shown here in two columns for space and readability reasons; PSpice will not accept two columns in a file to be analyzed.

```
*TRAN5.CIR—CASCADED AMPLIFIER
VGEN 1 0 AC 1M
*VGEN 1 0 SIN ( 0 1M 10K )
VSUPPLY 15 0 15
RGEN 1 2 10
CIN1 2 3 10U
R1-1 15 3 40K
R2-1 3 0 6.2K
Q1 4 3 5 Q2N2222A
RCOL-1 15 4 3K
RSWAMP-1 5 6 1K
REMIT-1 6 0 600
CBYPASS-1 6 0 100U
RCOL-2 8 0 3K
RSWAMP-2 10 9 1K
REMIT-2 15 10 600
CBYPASS-2 15 10 100U
CIN-3 8 11 10U
R1-3 15 11 30K
R2-3 11 0 30K
Q3 15 11 13 Q2N2222A
REMIT-3 13 0 1K
COUT 13 14 10U
RLOAD 14 0 100
.AC LIN 1 10K 10K
```

```
CIN-2 4 7 10U
R1-2 7 0 40K
R2-2 15 7 6.2K
Q2 8 7 9 Q2N2907A
.PRINT AC V(1) V(3) V(7) V(11) V(14)
*.TRAN 10U .5M 0 1U
*.PROBE
.END
```

FREQ	V(1)	V(3)	V(7)	V(11)	V(14)
1.000E+04	1.000E−03	9.981E−04	1.837E−03	3.910E−03	3.735E−03

6.6 Power Amplifiers

Because of the nonlinear characteristics of semiconductor devices, a complete mathematical description of transistor behavior is beyond the computational skills of most technicians. In general, this is not a problem. The nonlinear effects can often be ignored, greatly simplifying the mathematics and causing only minimal error. For example, technicians assume that .7 Volts are dropped by a silicon diode. While this is hardly ever true, it is a reasonable approximation.

For small-signal amplifiers, nonlinear effects are almost nonexistent because of the small portion of the transistor transconductance curve used. For power amplifiers or large signal amplifiers, the nonlinear effects become pronounced and deserve special consideration.

At the beginning of this chapter, we develop two DC load lines to describe the biasing needed for amplifier operation. Either load line was considered adequate as long as the signal was small and the biasing was near the middle of the line. For large signals, neither line adequately describes the AC characteristics of the amplifier, and a new line must be determined. The AC load line described in most textbooks takes into account the resistance of the load. Example 6.8 is exactly like Example 6.3 except that the load resistor has been reduced to 2 kilohms to make the effect of loading more pronounced.

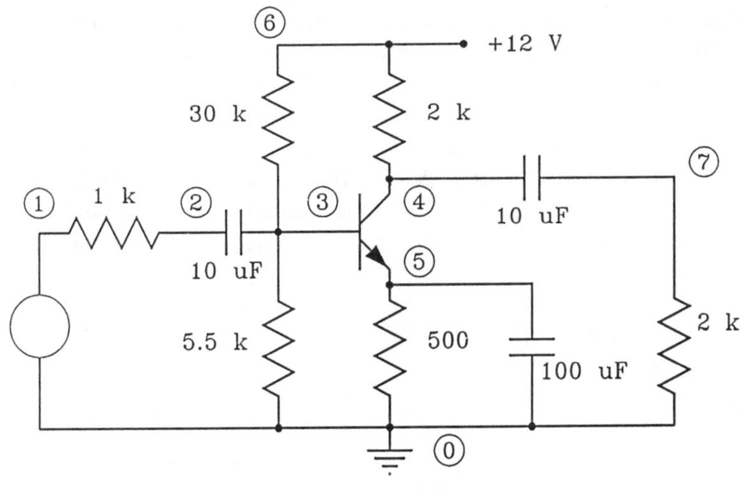

Example 6.8

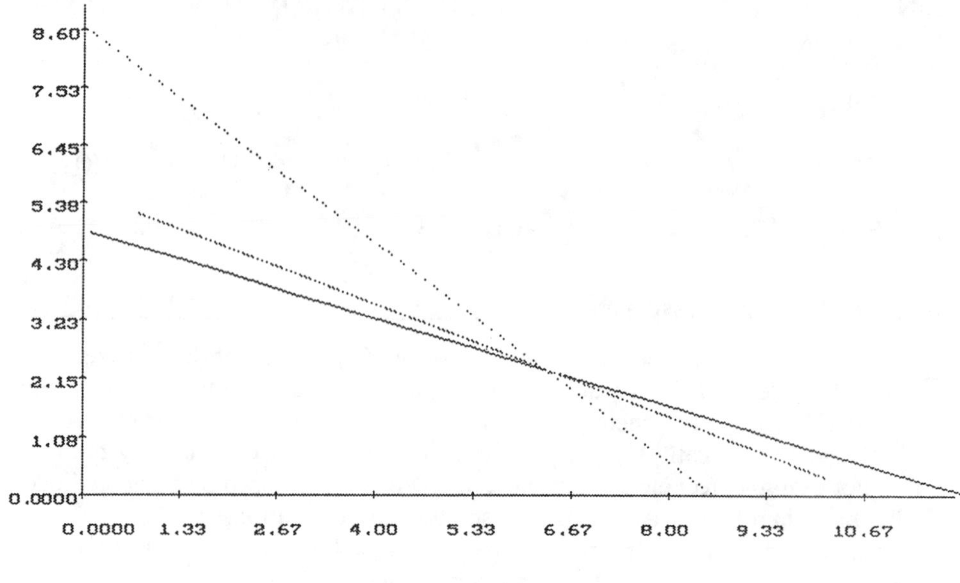

Figure 6.6

In this example, standard calculations result in an AC load line whose end points are (0, 8.6) and (8.6, 0). This line represents the locus of transistor voltage and current values possible for an AC waveform at a given biasing voltage. The graph in Figure 6.6 adds the AC load line to the DC load lines already developed.

These traces show that the quiescent point for the circuit lies at 6.36 Volts with a current of 2.25 milliamps. Since the AC load line intercepts the X-axis at 8.62 Volts, most textbooks consider the circuit to have a maximum peak-to-peak output amplitude of 4.5 Vp-p or $2 \times (8.62 - 6.36)$. The implication is that for output smaller than 4.5 Vp-p the amplifier will work properly, and for output larger than 4.5 Vp-p the wave will be clipped. This is an example of a technician's textbook being overly simplified and misleading. PSpice can show that the amplifier neither works well for output less than 4.5 Vp-p nor clips at voltages moderately larger.

Example 6.8 is designed to be analyzed twice. The first analysis uses a 25 millivolt input signal and the second uses a 75 millivolt signal. Calculations indicate that the amplifier has a theoretical gain of 82.6. This indicates that the peak-to-peak output signals should be 2.46 Volts and 7.38 Volts respectively, with the second output clipped at 4.5 Volts.

```
*TRAN6.CIR—POWER AMPLIFIER
.INC \SPICE\TRAN.LIB
VSUPPLY 6 0 12
CIN 2 3 10U
COUT 4 7 10U
CBYPASS 5 0 100U
```

```
RIN 1 2 1K
R1 6 3 30K
R2 3 0 5.5K
RCOL 6 4 2K
REMIT 5 0 500
RLOAD 7 0 2K
Q1 4 3 5 Q2N2222A
VIN 1 0 SIN(0 25M 10K)
*VIN 10 SIN(0 75M 10K)
.TRAN 10U .5M 0 1U
.PROBE
.END
```

If PROBE is used to examine the voltage at node 7, the output from the first analysis, shown in Figure 6.7, indicates that the signal has been distorted by the amplifier. Instead of the expected −1.23 Volts to +1.23 Volts, the waveform is compressed in the positive region and elongated in the negative region.

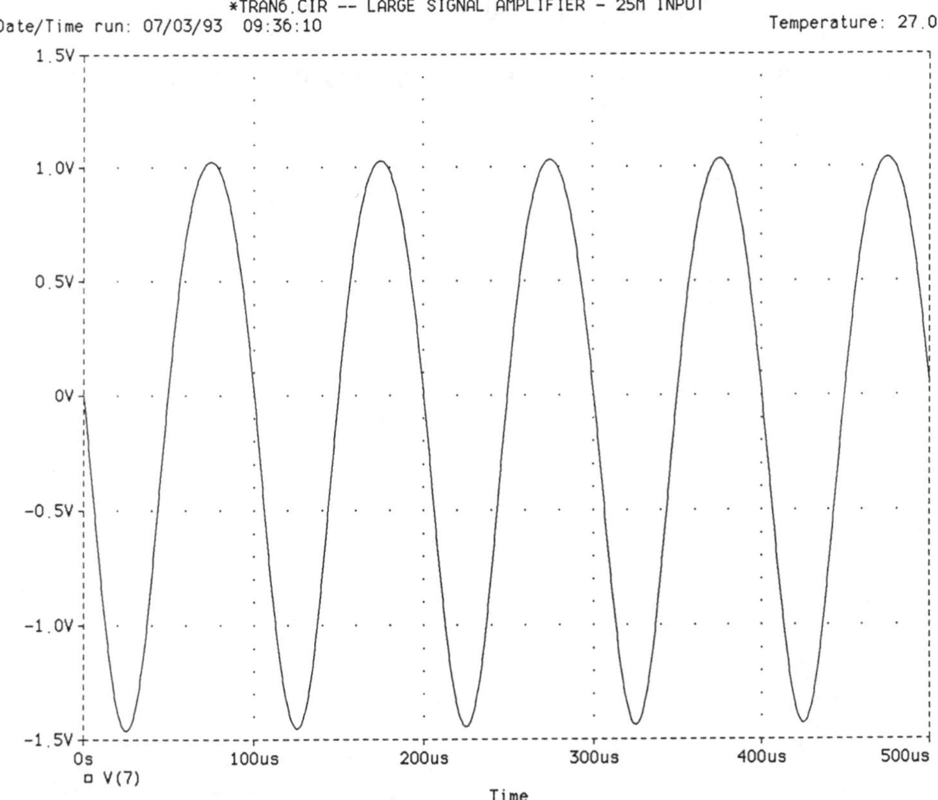

Figure 6.7

Choosing the "X-axis" option from the main menu and performing a Fourier analysis makes it apparent that the output contains a DC component and a second harmonic, as shown in Figure 6.8.

The second analysis, using the 75 millivolt input, shows even greater distortion but no clipping. See Figure 6.9.

The Fourier analysis for the second output, shown in Figure 6.10, indicates that the DC component and the second harmonic have grown larger relative to the amplitude of the fundamental, and higher harmonics have been added to the output.

The problem of large signal distortion in class A amplifiers cannot be avoided because transistors are inherently nonlinear devices. However, the effect can be lessened by changing the biasing so the Q-point is in the center of the AC load line rather than the DC load line.

6.7 Class B Amplifiers

All the amplifiers described in this chapter so far are class A amplifiers. This means that the transistor conducts current through the entire cycle of the input signal. While class A operation is the simplest and most common form of amplifier operation, it suffers from low efficiency, high distortion, and an inadequate ability to drive heavy loads.

Class B amplifiers are biased at (or near) the cutoff point on the load line. This

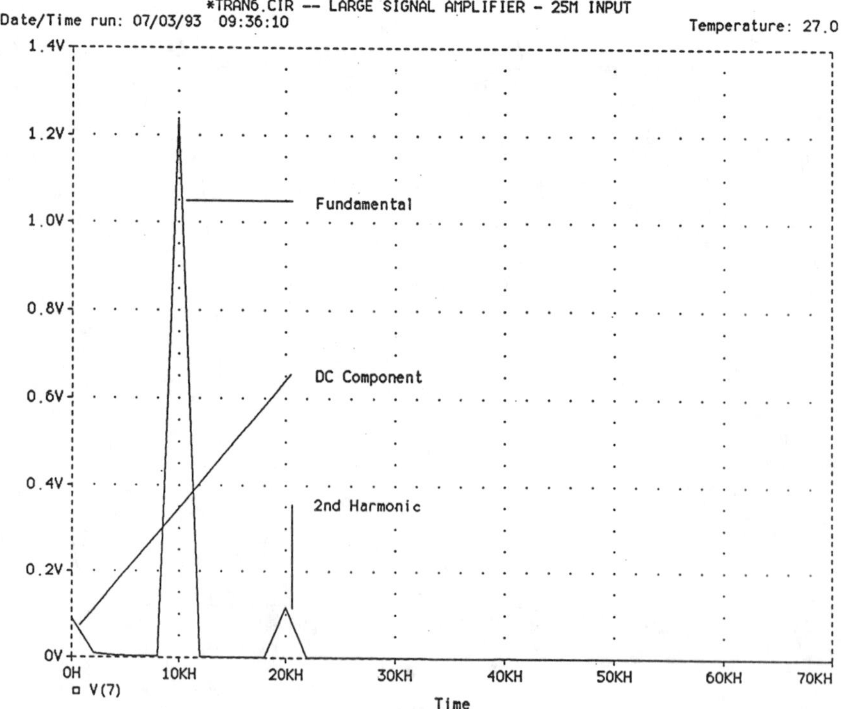

Figure 6.8

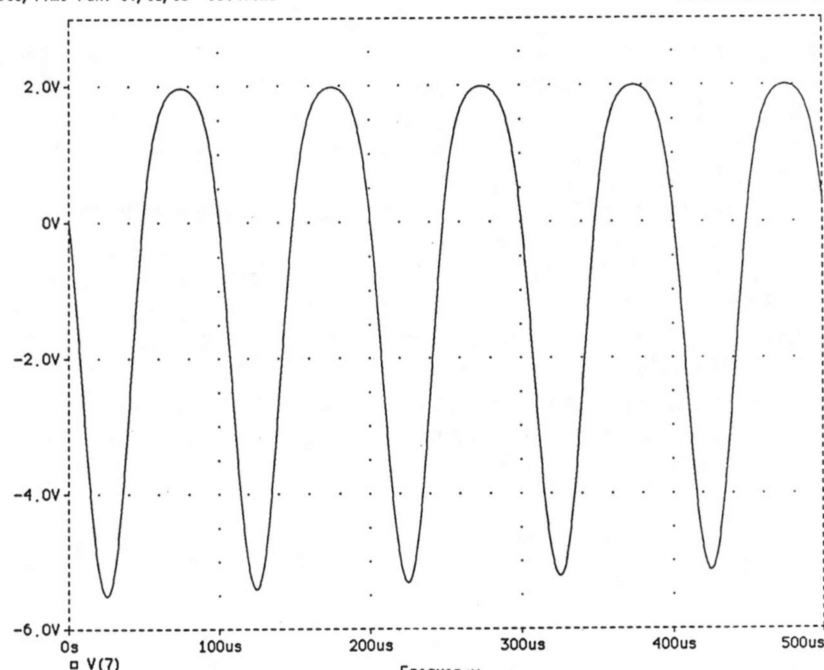

Figure 6.9

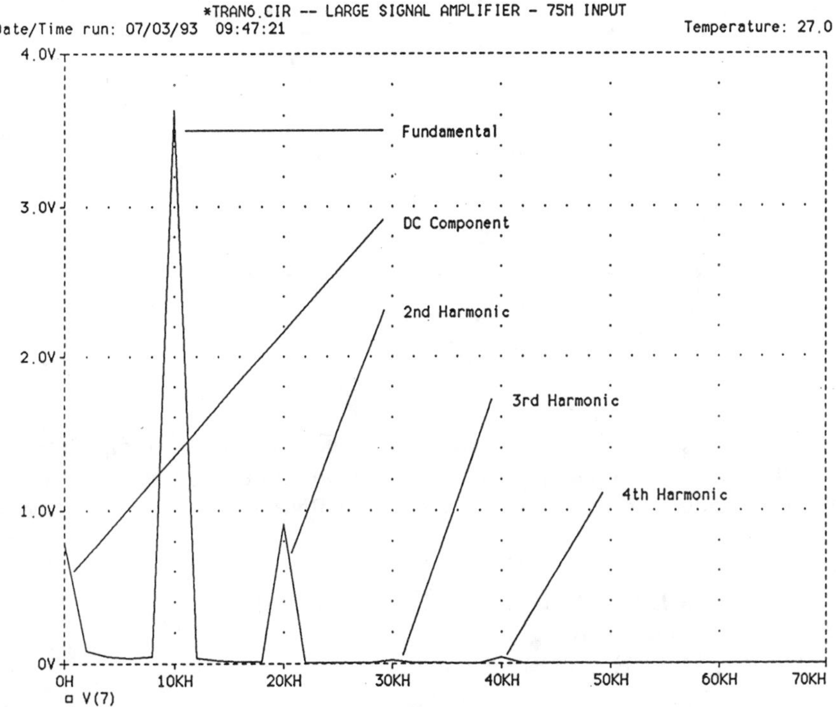

Figure 6.10

implies that the transistor conducts current for only one-half the cycle of the input signal. This allows the amplifier to output a large, relatively undistorted signal that is rectified. To regain the other half of the signal, a second transistor is used that conducts only while the first transistor is in cutoff. Since only one transistor is conducting at any given time, this arrangement is called push–pull.

Class B push–pull amplifiers are usually designed around a matched pair of transistors. This means that two transistors with nearly identical characteristics, except that one is an NPN and the other is a PNP, are used. The evaluation library contains the 3904/3906 matched pair. Ideally the diodes used in biasing the transistors should have characteristics identical to the base-emitter junction of the transistors. For our purposes, the 1N4148 proves to be close enough.

Example 6.9 illustrates the basic idea of providing the same input signal to two transistors. The upper transistor conducts on the positive half-cycle and the lower transistor on the negative half-cycle.

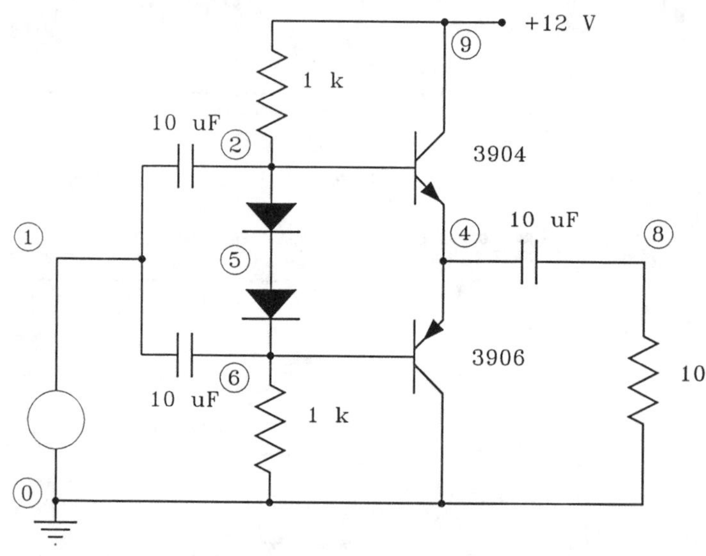

Example 6.9

Only two items are worth reemphasizing in the circuit description. Both TRAN. LIB and DIODE.LIB must be included in the circuit description. The order of the connecting nodes for both the NPN and the PNP is the same—i.e., collector-base-emitter.

```
*TRAN7.CIR—CLASS B PUSH-PULL AMPLIFIER
.INC \SPICE\TRAN.LIB
.INC \SPICE\DIODE.LIB
VSUPPLY 9 0 12
```

```
VSIGNAL 1 0 SIN(0 4 10K)
CIN1 1 2 10U
CIN2 1 6 10U
COUT 4 8 10U
R1 9 2 1K
R2 6 0 1K
D1 2 5 D1N4148
D2 5 6 D1N4148
Q1 9 2 4 Q2N3904
Q2 0 6 4 Q2N3906
RLOAD 8 0 10
.TRAN 10U 1M .5M 1U
.PROBE
.END
```

Using PROBE to view the input and output waveforms (see Figure 6.11) shows that the class B amplifier can handle an 8 Volt peak-to-peak signal without adding any

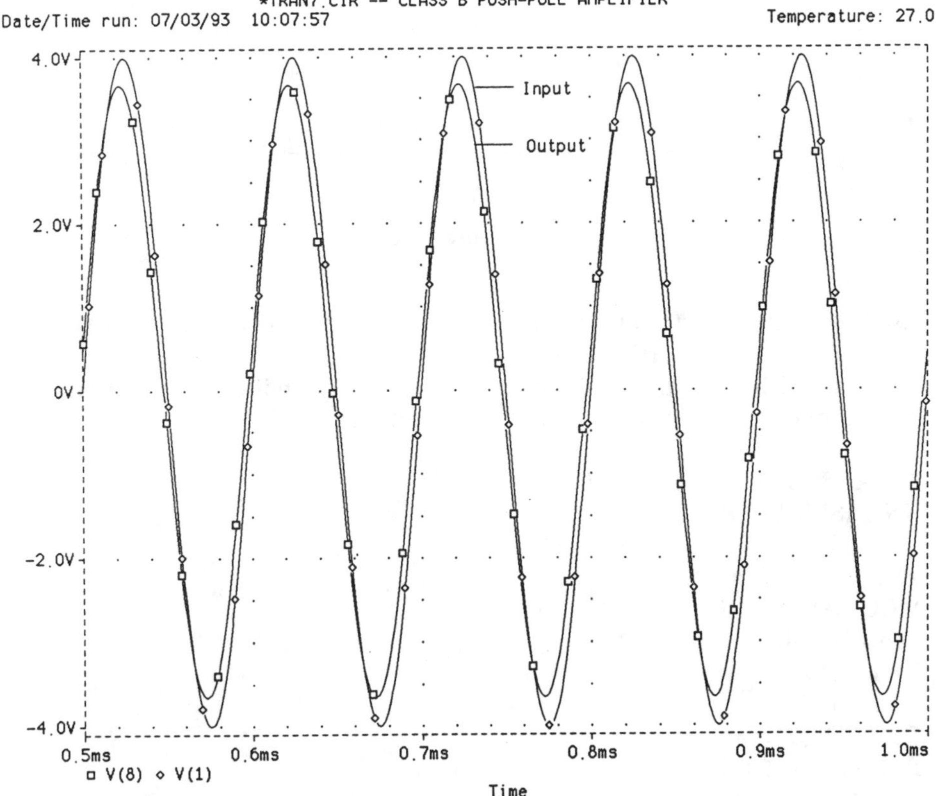

Figure 6.11

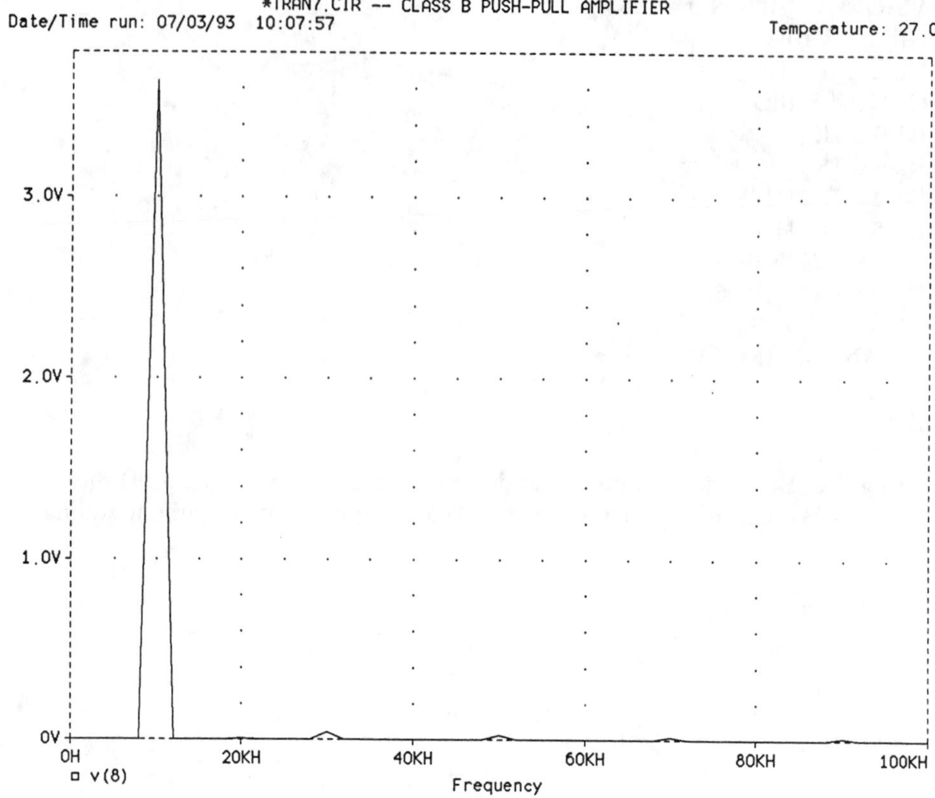

Figure 6.12

visible distortion. Performing a Fourier analysis, as shown in Figure 6.12, confirms that the harmonic distortion is minimal.

Example 6.10 is a more practical class B push-pull amplifier. A class A amplifier is used to drive current through the biasing network of the class B.

```
*TRAN8.CIR–CLASS B WITH PREAMP
.INC \SPICE\TRAN.LIB
.INC \SPICE\DIODE.LIB
VSUPPLY 1 0 20
VSIGNAL 8 0 SIN(0 .75 10K)
CIN 8 7 100U
R1 1 7 20K
R2 7 0 1.9K
Q1 4 7 9 Q2N3904
REMIT 9 0 100
D1 2 6 D1N4148
```

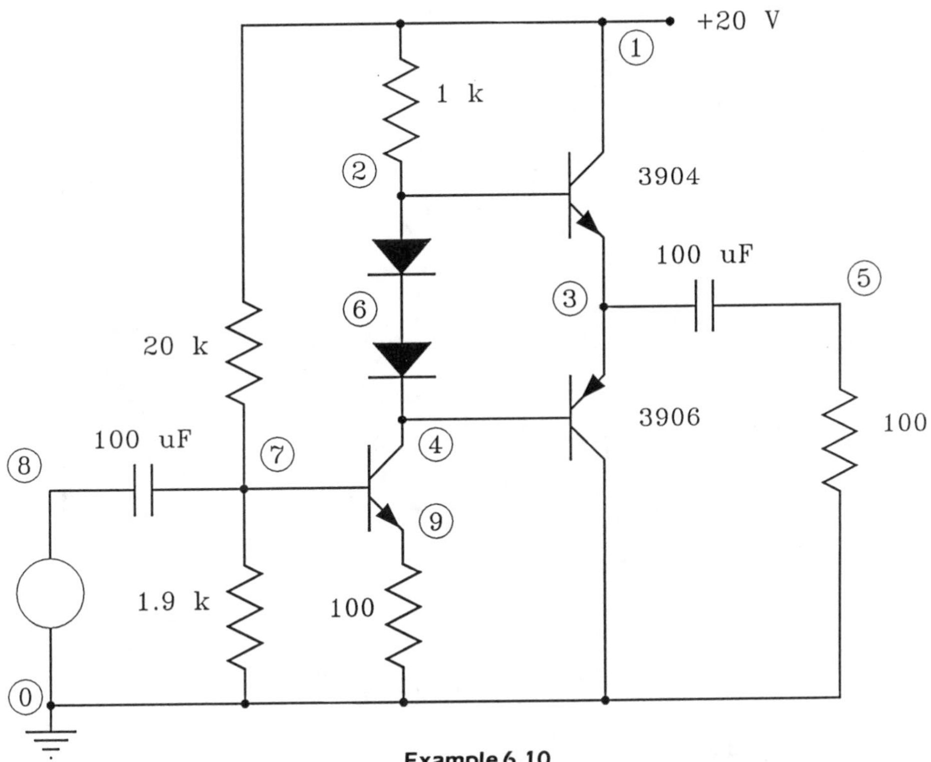

Example 6.10

```
D2 6 4 D1N4148
RCOL 1 2 1K
Q2 1 2 3 Q2N3904
Q3 0 4 3 Q2N3906
COUT 3 5 100U
RLOAD 5 0 100
.TRAN 10U 1M .5M 1U
.PROBE
.END
```

A visual inspection of the signal voltage developed across the load reveals minimal distortion, as shown in Figure 6.13. The 11 Volt peak-to-peak signal is slightly compressed in the positive region and slightly elongated in the negative region. However, considering the size of the waveform, it is significantly less distorted than much smaller signals developed with class A power amplifiers.

From the Fourier analysis of the signals at node 4 and node 5, shown in Figure 6.14, it is clear that the distortion present at the load is due to the class A amplifier, and the push–pull amplifier has added no appreciable distortion.

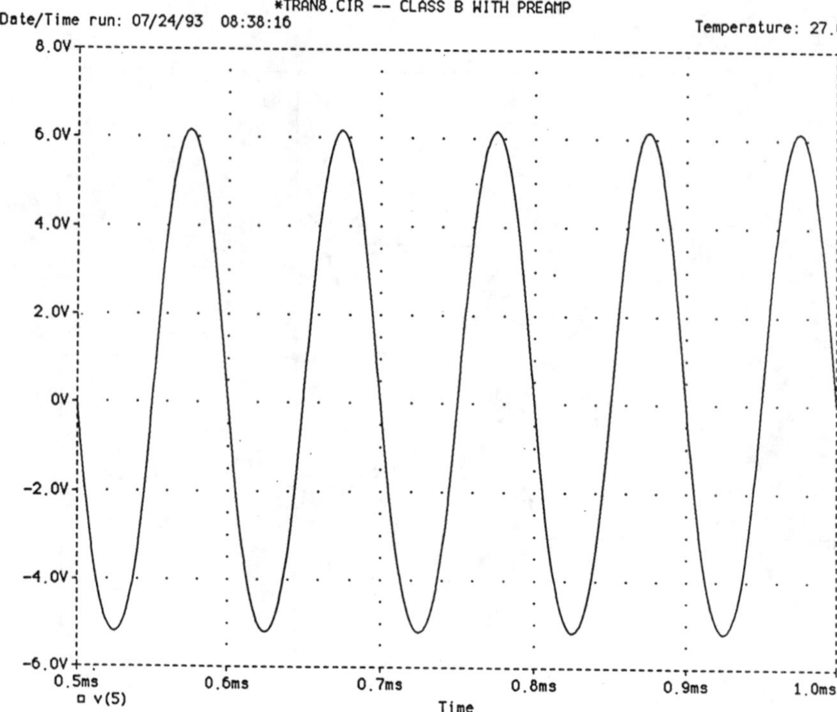

Figure 6.13

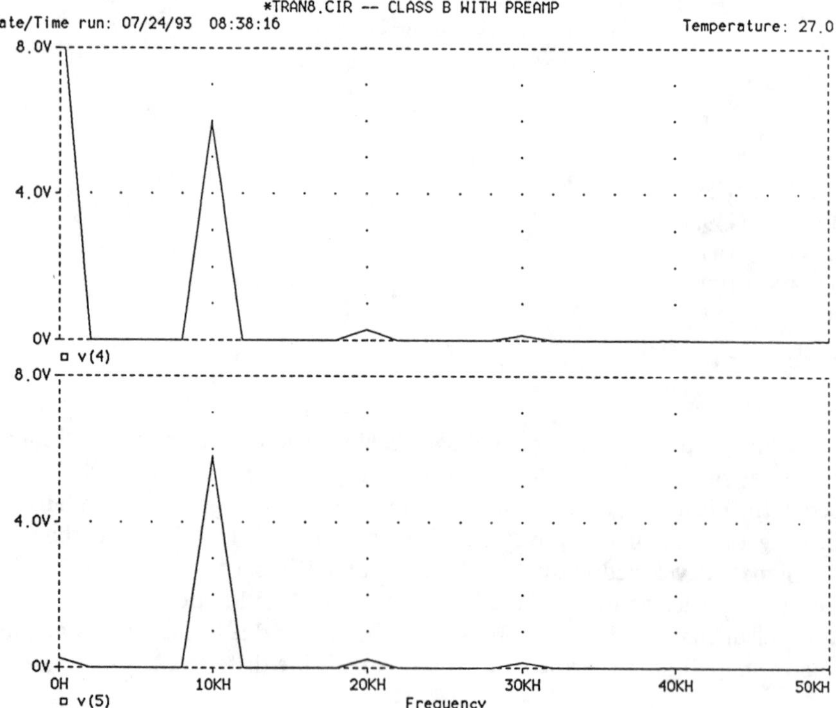

Figure 6.14

7

Field-Effect Transistor Circuits

In the most general terms, field-effect transistors perform the same functions as bipolar junction transistors. Both are used in building amplifiers, switches, and integrated circuit chips. A slightly closer look reveals not only that they are quite different in construction but also that their characteristics differ greatly. Field-effect transistors provide much less voltage gain but have far greater input impedance. Using PSpice simulations, we can demonstrate typical gains and show the effect of very large input impedances. PSpice can also illustrate more subtle differences such as output distortion effects.

Field-effect transistors (FETs) can be roughly divided into junction field-effect transistors (JFETs) and metal oxide semiconductor field-effect transistors (MOSFETs). In bipolar transistor amplifiers, the signal usually enters the transistor at the base and sees a diode that has been forward biased. In a JFET amplifier, the signal usually enters at the gate and sees a reverse-biased diode. This results in an input impedance measured in tens or hundreds of Megohms. A MOSFET has a glass (SiO_2) insulator on the gate and an input impedance measured in tens or hundreds of gigohms.

The evaluation library contains two N-channel JFETs, one N-type MOSFET, and one P-type MOSFET. As in previous chapters, it is necessary to create a FET library if access to the library models is to be made by a .INC statement. Copy EVAL.LIB into FET.LIB and delete everything except the following.

```
.model J2N3819  NJF(Beta=1.304m Betatce=-.5 Rd=1 Rs=1 Lambda=2.25m
+       Vto=-3
+       Vtotc=-2.5m Is=33.57f Isr=322.4f N=1 Nr=2 Xti=3 Alpha=311.7
+       Vk=243.6 Cgd=1.6p M=.3622 Pb=1 Fc=.5 Cgs=2.414p Kf=9.882E-18
+       Af=1)
*       National     pid=50         case=TO92
*       88-08-01 rmn     BVmin=25

.model J2N4393  NJF(Beta=9.109m Betatce=-.5 Rd=1 Rs=1 Lambda=6m
+       Vto=-1.422
+       Vtotc=-2.5m Is=205.2f Isr=1.988p N=1 Nr=2 Xti=3 Alpha=20.98u
```

```
+       Vk=123.7 Cgd=4.57p M=.4069 Pb=1 Fc=.5 Cgs=4.06p Kf=123E-18
+       Af=1)
*       National      pid=51        case=TO18
*       88-07-13 bam      BVmin=40

.model IRF150   NMOS(Level=3 Gamma=0 Delta=0 Eta=0 Theta=0 Kappa=0
+       Vmax=0 Xj=0
+       Tox=100n Uo=600 Phi=.6 Rs=1.624m Kp=20.53u W=.3 L=2u Vto=2.831
+       Rd=1.031m Rds=444.4K Cbd=3.229n Pb=.8 Mj=.5 Fc=.5 Cgso=9.027n
+       Cgdo=1.679n Rg=13.89 Is=194E-18 N=1 Tt=288n)
*       Int'l Rectifier pid=IRFC150 case=TO3
*       88-08-25 bam      creation

.model IRF9140  PMOS(Level=3 Gamma=0 Delta=0 Eta=0 Theta=0 Kappa=0
+       Vmax=0 Xj=0
+       Tox=100n Uo=300 Phi=.6 Rs=70.6m Kp=10.15u W=1.9 L=2u Vto=-3.67
+       Rd=60.66m Rds=444.4K Cbd=2.141n Pb=.8 Mj=.5 Fc=.5 Cgso=877.2p
+       Cgdo=369.3p Rg=.811 Is=52.23E-18 N=2 Tt=140n)
*       Int'l Rectifier pid=IRFC9140 case=TO3
*       88-08-25 bam creation
```

7.1 Transconductance Curve

A bipolar transistor can be thought of as a switch that is normally open and will not allow current to flow from collector to emitter until a positive voltage at the base provides the proper bias to close it. A JFET, which usually conducts current from drain to source, can be thought of as a normally closed switch. To be useful, it must have a negative voltage from the gate to the source to turn off the switch.

In analog circuits, we generally avoid having a transistor completely closed or completely open. More commonly, the switches are biased partially open. Since the idea of a switch that can be partially open is not very intuitive, an analogy is often made to a water valve. A BJT operates like a valve in which the water flows from collector to emitter, and the amount of water is controlled by the voltage at the base, which acts as the handle of the valve. The positive-biasing voltage opens the valve halfway, and the AC voltage allows either more or less water to travel from the collector end to the emitter end. For a JFET the valve is normally open, and a biasing voltage is needed at the gate to partially close off the water moving from drain to source.

The amount of current that can pass from drain to source for any given gate-to-source voltage is shown by the transconductance curve. Clearly the drain current is maximum when the gate-to-source voltage is zero and the switch is totally closed. It is equally apparent that there must be some negative voltage to open the switch, stopping all current. These are the end points of the transconductance curve. Proper biasing requires a gate-to-source voltage somewhere in between.

The transconductance curve can be developed by the test circuit shown in Example 7.1.

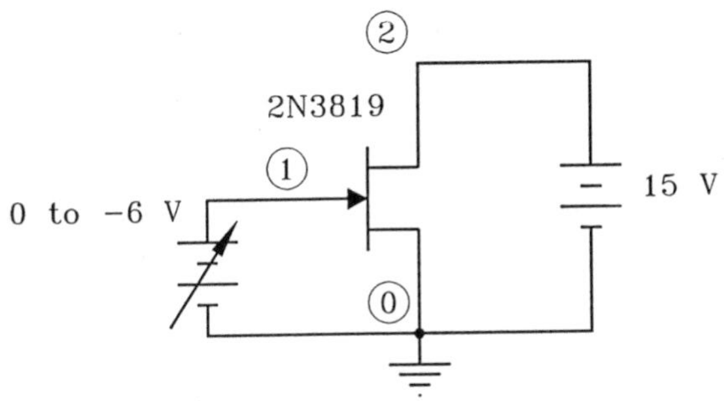

Example 7.1

```
*FET1.CIR—FET TRANSCONDUCTANCE CURVE
.INC \SPICE\FET.LIB
VDRAIN 2 0 15
VGATE 1 0 -6
J1 2 1 0 J2N3819
.DC VGATE -6 0 .1
.PROBE
.END
```

The only new feature in the circuit description is the description of the JFET. The general form for a JFET is

J<name> <drain> <gate> <source> <model>

The included library contains the model for the 2N3819 JFET. The .DC analysis sweeps the gate voltage from −6 Volts to 0 Volts in .1 Volt increments.

For the transconductance curve, the gate voltage sweep is the independent variable. The value of interest is the drain current as a function of gate–source voltage. In PROBE, drain current can be plotted as ID(J1). Notice that there is no need to resort to adding a fictional zero valued device to the circuit to obtain the drain current. Unlike .PRINT statements, which allow only current values through two-terminal components, PROBE accepts requests for current entering and leaving three-terminal devices. After adjusting both axes, PROBE displays the graph shown in Figure 7.1. The transconductance curve indicates that for the 2N3819 JFET a gate-to-source voltage of −1 Volts will bias the transistor between the saturation and cutoff extremes, slightly shifted toward the more linear portion of the plot.

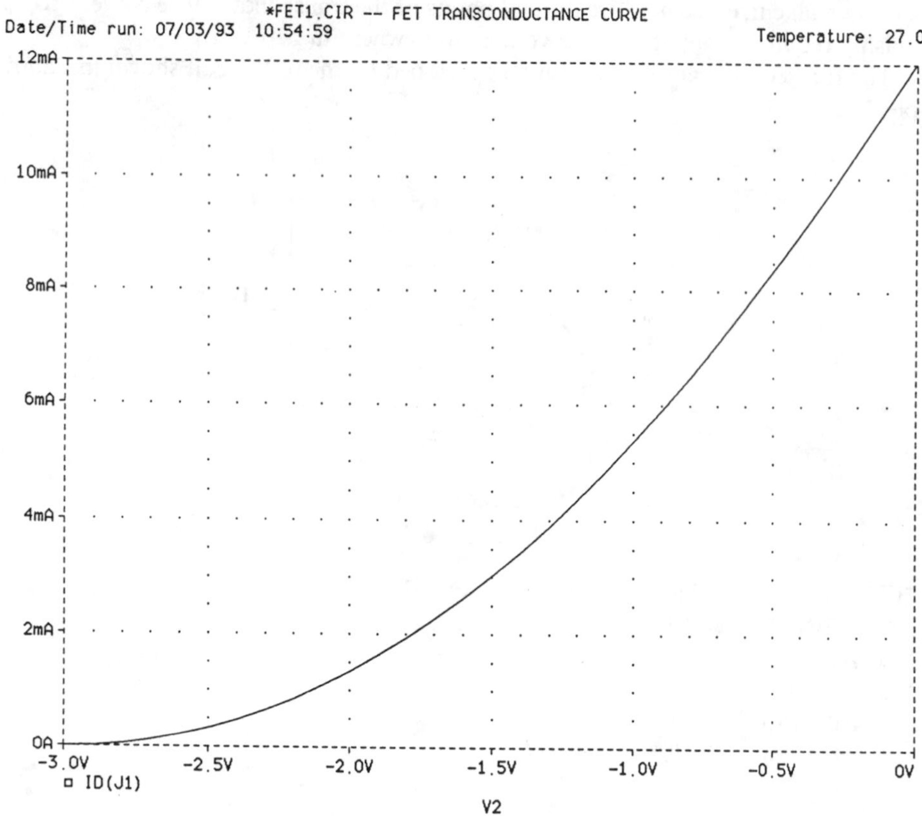

Figure 7.1

7.2 Drain Current Curves

The transconductance curve is generated by examining the drain current as the drain voltage is held steady and the gate-to-source voltage is varied. A useful family of curves can be developed by fixing the gate-to-source voltage at evenly spaced levels and varying the drain voltage for each level. These curves indicate the ohmic, active, and breakdown regions for each gate-to-source voltage.

PSpice can provide a family of curves by using a sweep of DC voltages for the drain voltage and the .STEP function to change the gate voltage. This causes the sweep analysis to be performed for each gate voltage in the step range. See Example 7.2.

```
*FET2.CIR—FET DRAIN CURRENT CURVES
.INC \SPICE\FET.LIB
VDRAIN 2 0 15
.PARAM VOLTAGE=1
.STEP PARAM VOLTAGE −3 0 .5
```

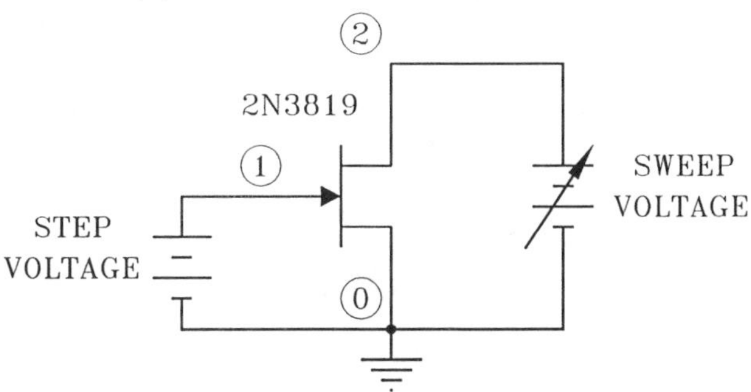

Example 7.2

```
VGATE 1 0 {VOLTAGE}
J1 2 1 0 J2N3819
.DC V1 0 40 1
.PROBE
.END
```

.PARAM and .STEP in this circuit description are used much as the potentiometer descriptions are used in previous examples. .PARAM defines a variable called "VOLTAGE" and gives it an initial value of 1. .STEP causes "VOLTAGE" to vary its value from −3 to 0 in steps of one-half volt. At each step, the .DC analysis is performed from 0 to 40 volts in one-volt increments. In other words, the analysis is performed seven times; once for each of the gate voltages 0, .5, 1, 1.5, . . ., 3.

When the analysis is complete and the PROBE program runs, the user is given the option to graph "All_Dc_sweep" or to "Select_sections." "All_Dc_sweep" graphs the entire family of seven curves. "Select_sections" gives the user the opportunity to choose specific curves based on selected gate voltage values. The selections can be made by changing the highlight position with the arrow keys and hitting the space bar or, more conveniently, by pointing to a selection with a mouse and clicking the left button.

Choosing "All_Dc_sweep" and requesting the drain current, ID(J1), results in the graph shown in Figure 7.2.

This graph adequately illustrates both the active and the breakdown regions of the transistor response, but it compresses the ohmic region. Changing the scale of the X-axis spreads the ohmic region but produces a rather crude graph because points were chosen at 1 Volt intervals. A more satisfactory graph can be made by altering the circuit description, using a smaller range and smaller step for the .DC sweep (.DC 0 5 .2). Rerunning PSpice yields the graph shown in Figure 7.3.

7.3 Small-Signal Amplifiers

The transconductance curve contains needed information for the proper biasing of a JFET amplifier. Adding an input signal, a load, and the necessary coupling and

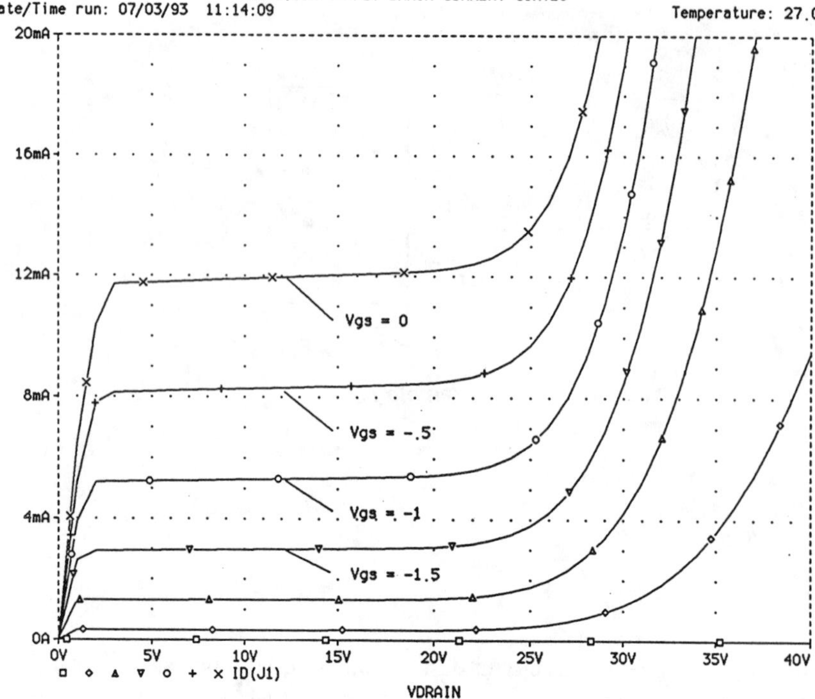

Figure 7.2

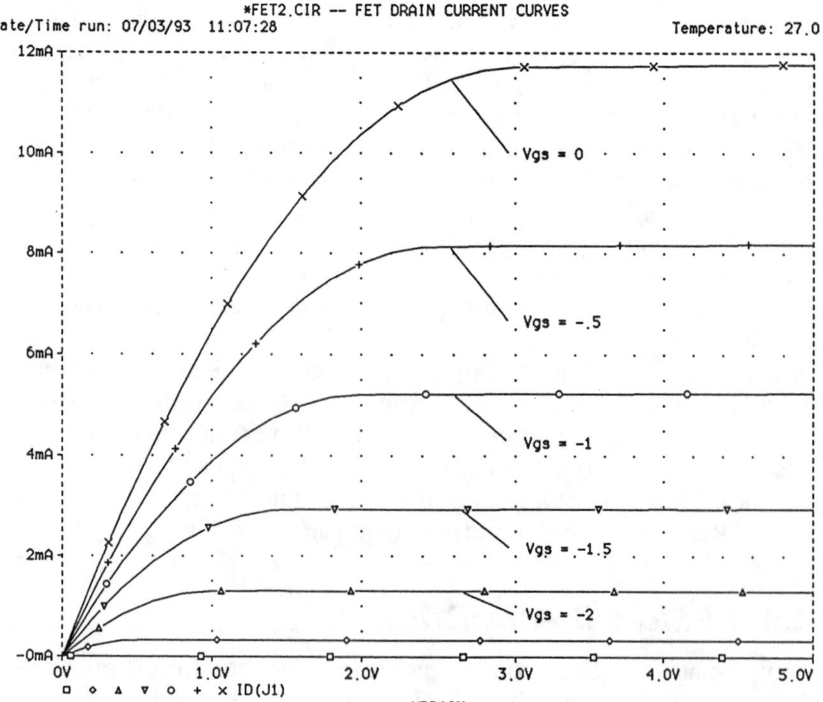

Figure 7.3

bypass capacitors completes the amplifier circuit. Example 7.3 is a typical self-biased amplifier. The 1 MΩ resistor provides a ground-level DC biasing voltage at the gate. The current through the source resistor causes node 5 to have a positive DC voltage, effectively reverse biasing the gate–source junction.

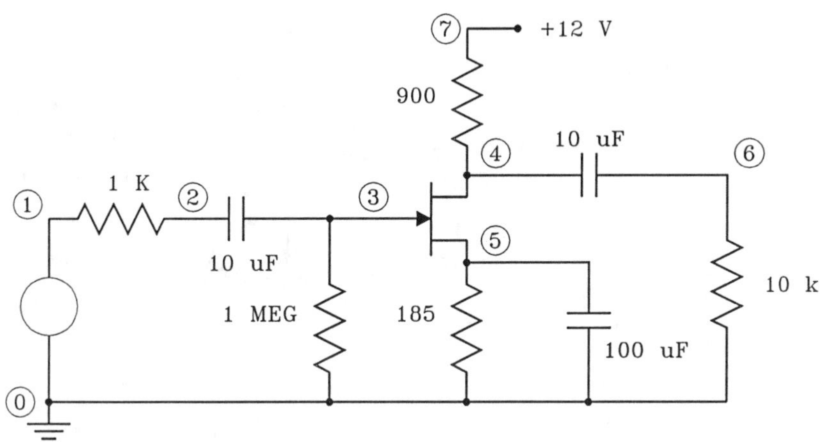

Example 7.3

JFET amplifiers are subject to distortions similar to those associated with large-signal operation of bipolar transistors. Consequently, Example 7.3 uses a 1 millivolt input signal leaving large-signal operation to be examined in section 7.4. The initial analysis for the amplifier is time based. The circuit description is then altered to allow for bandwidth examination and inspection of AC and DC numeric values.

```
*FET3.CIR—SMALL SIGNAL AMPLIFIER
.INC \SPICE\FET.LIB
VSUPPLY 7 0 12
RGEN 1 2 2K
CIN 2 3 10U
RBIAS 3 0 1MEG
J1 4 3 5 J2N3819
RSOURCE 5 0 185
CSOURCE 5 0 100U
RDRAIN 7 4 900
COUT 4 6 10U
RLOAD 6 0 10K
****
VGEN 1 0 SIN (0 1M 10K)
.TRAN 10U .5M 0 1U
.PROBE
****
.END
```

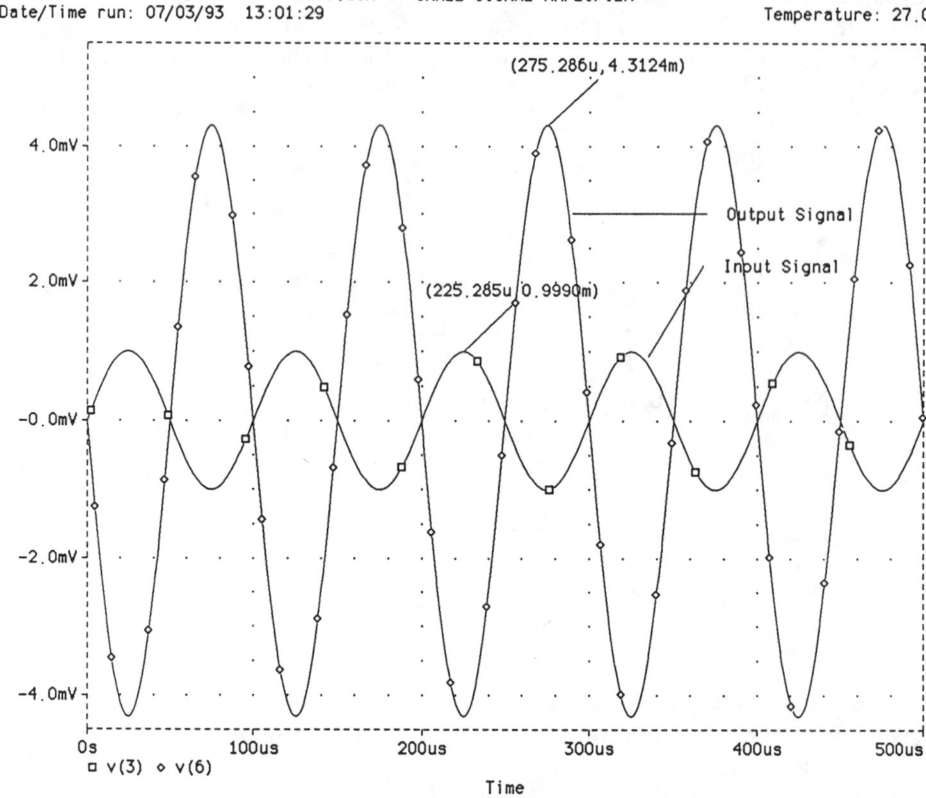

Figure 7.4

The generator signal is a 1 millivolt, 10 kilohertz signal that is analyzed for the first five input cycles. The graph shown in Figure 7.4 indicates that the common-source JFET amplifier inverts the signal just as the common-emitter amplifier but provides much less gain.

Changing the lines between the comment line markers to

```
VGEN 1 0 AC 1M
.AC DEC 100 1 10G
.PROBE
```

provides a frequency sweep analysis and the bandwidth information shown in Figure 7.5.

Finding numeric values for AC voltages requires a frequency sweep of a single frequency. The lines between the markers become

```
VGEN 1 0 AC 1M
```

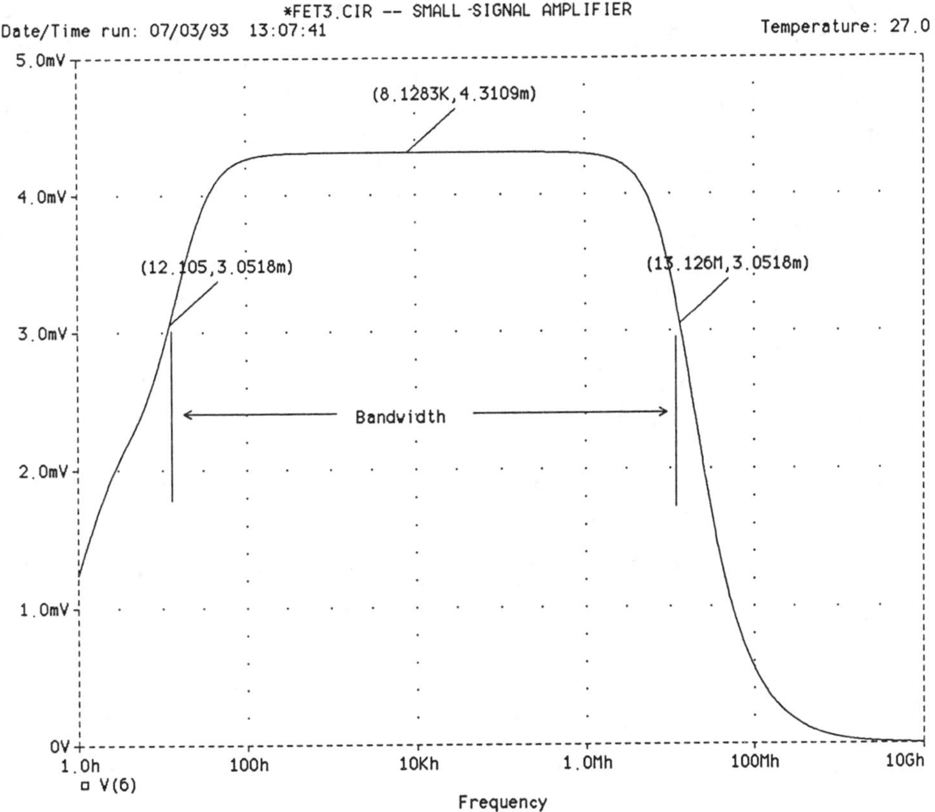

Figure 7.5

.AC LIN 1 10K 10K
.PRINT AC VM(1) VM(3) VM(6) VP(3) VP(6)

Deleting the noninteresting information in the .OUT file leaves

FREQ	VM(1)	VM(3)	VM(6)	VP(3)	VP(6)
1.000E+04	1.000E−03	9.980E−04	4.311E−03	−4.193E−02	−1.800E+02

The gain and the input impedance for the amplifier can be calculated from these data. The gain of approximately 4.3 [VM(6)/VM(3)] is typical for JFET amplifiers. The input impedance requires a little more calculating. Inserting the source voltage, the gate voltage, and the generator output impedance into the voltage divider formula shows the amplifier input impedance to be

$$\mathbf{.998/1.000} = x/(\mathbf{2k}+\mathbf{x})$$
$$x = \mathbf{998k}$$

Since this value is close to the 1 MΩ value of the biasing resistor, the parallel input impedance of the JFET must be very large. The exact value of the input impedance of the transistor could be calculated, but it would have little meaning. Extremely small changes in the gate voltage produce very large changes in the calculated transistor impedance. Accuracy is limited by the number of significant figures in the voltage data. The best that can be inferred from the PSpice numbers is that the input impedance to the JFET is between 5 and 500 MΩ.

DC biasing levels are obtained by performing a .DC analysis. Inserting the following lines between the markers in the circuit description tells PSpice to provide the voltages at the drain, gate, and source, and across the gate–source diode.

```
VGEN 1 0 SIN (0 1M 10K)
.DC VSUPPLY 12 12 1
.PRINT DC V(4) V(3) V(5) V(3,5)
```

The information in the .OUT file indicates that the gate–source voltage, V(3,5), is close to −1 Volts. Examining the transconductance curve shows this is an acceptable value for V_{GS}. Plotting the drain voltage of 7.2 Volts on the drain current curves shows the transistor is operating in the active region. The voltage at node 3 is slightly over 1 microvolt and justifies the assumption that the 1 MΩ resistor effectively ties the gate to ground.

VSUPPLY	V(4)	V(3)	V(5)	V(3,5)
1.200E+01	7.201E+00	1.172E−06	9.865E−01	−9.865E−01

7.4 Large-Signal Amplifiers

When an input signal is added to the biasing voltage of a FET amplifier, the gate–source voltage for the amplifier travels along the transconductance curve above and below the Q-point. Since the gain of the amplifier is proportional to the slope of the curve at the gate–source voltage point (V_{GS}), the gain is larger when the input is positive than it is when the input is negative. The inverting action of the common source amplifier makes the negative half-cycle of the output larger in magnitude than the positive half-cycle.

If the input signal is small, V_{GS} remains close to the Q-point where the transconductance curve is virtually linear. A linear transconductance curve implies that both half-cycles of the input receive the same amplification and the output signal is undistorted. If the input signal is large, a large range of the transconductance curve is used. The result is a significantly varying gain and a distorted output.

This distortion is similar, but not identical, to the distortion in a large-signal BJT amplifier. The transconductance curve of a JFET is a parabola or square law curve. The transconductance curve of a BJT is dependent upon the curve of a forward-biased diode. The result is somewhat less distortion in the JFET, and Fourier analysis will show that distortion in the output of a JFET amplifier is easier to filter than the output distortion from a BJT.

The amplifier circuit from the previous section is reused with just two modifications for a large-signal amplifier. The input amplitude is increased to 1.2 Volts for large-signal operation and the load is reduced to 2 kilohms for better comparison to the large-signal BJT amplifier analyzed in the previous chapter. An amplitude of 1.2 Volts was chosen to produce an output amplitude for the fundamental of approximately 3.7 Volts, again to allow comparison to the output for Example 6.6.

```
VGEN 1 0 SIN (0 1.2 10K)
RLOAD 6 0 2K
```

A visual inspection of the output waveform, shown in Figure 7.6, indicates less distortion in the JFET output. A better comparison can be made by examining the results of the Fourier analysis shown in Figure 7.7. The analysis of the JFET amplifier shows a smaller DC component and a smaller second harmonic. The higher harmonics are larger in the JFET output, but this is considered a "better" type of distortion. As the spectral distance between the desired signal and the unwanted harmonics becomes greater, the filtering of those harmonics becomes easier.

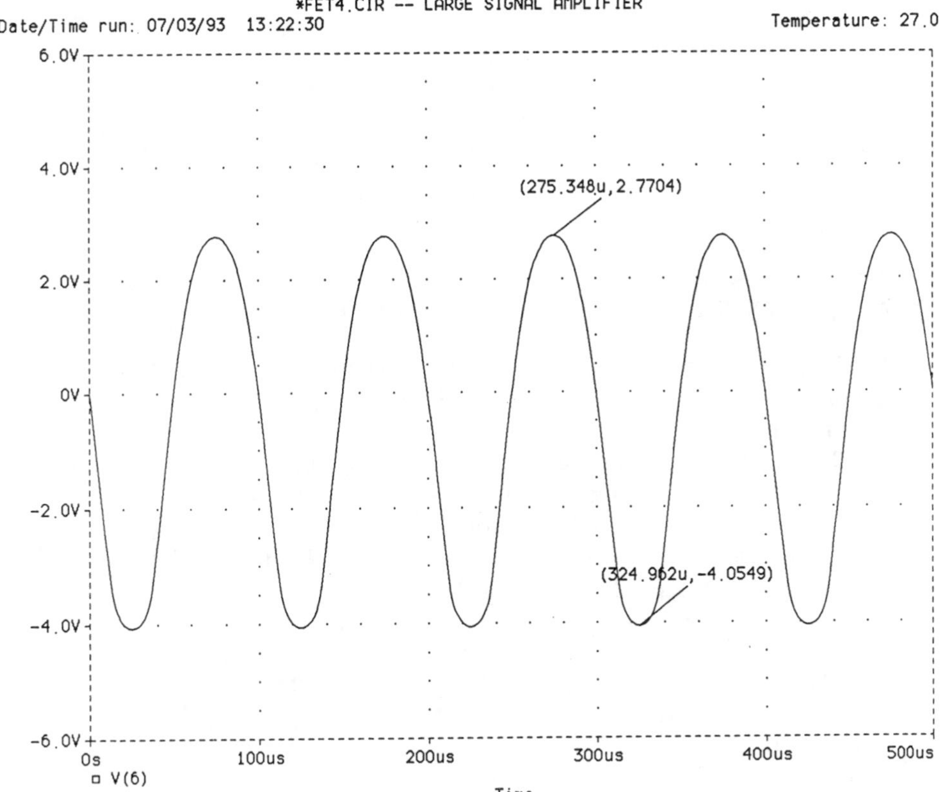

Figure 7.6

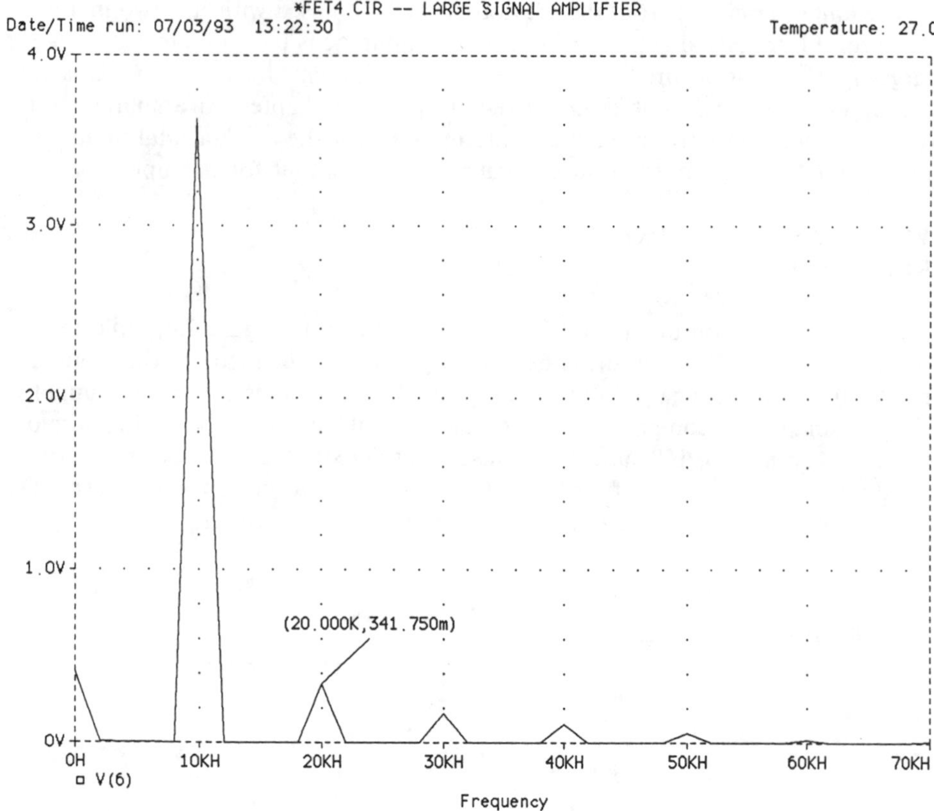

Figure 7.7

7.5 MOSFET Transconductance Curve

MOSFETs are named for the metal oxide insulator that covers the gate of the transistor. The insulator, usually glass (SiO_2), gives the MOSFET its extremely high input impedance. MOSFETs are most often used as switches, either in integrated circuit chips or in power applications. Analog applications are less common, but MOSFETs, particularly dual-gate MOSFETs, are used in mixers and in radios as insertion points for automatic gain control.

The PSpice evaluation library provides two MOSFETs. The IRF150 is an N-type E-MOSFET and the IRF9140 is a P-type D-MOSFET. Example 7.4 is a test circuit for developing the transconductance curve for the IRF150.

```
*FET5.CIR—MOSFET TRANSCONDUCTANCE CURVE
.INC \SPICE\FET.LIB
VSUPPLY 1 0 15
VIN 2 0 1
RDRAIN 1 11 10
```

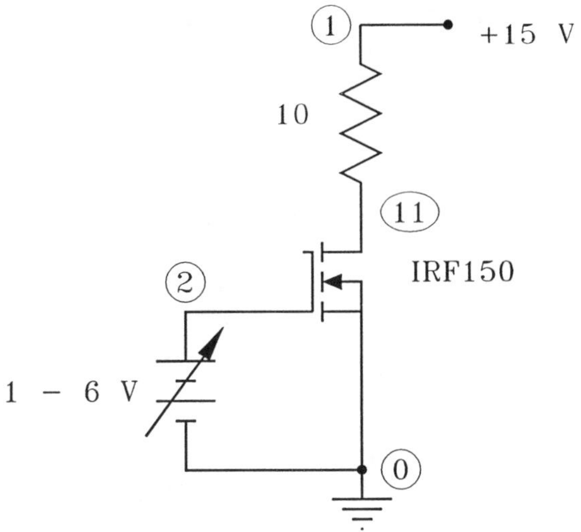

Example 7.4

```
M1 11 2 0 0 IRF150
.DC VIN 1 6 .1
.PROBE
.END
```

The general form for a MOSFET is

M<name> <drain> <gate> <source> <substrate> <model>

The transconductance curve in Figure 7.8 shows that the transistor does not begin to conduct until nearly three volts potential exists from gate to source. An increase of less than one volt saturates the transistor with a current of 1.5 Amps. This indicates that the transistor is better suited for power switching circuits than for amplifiers.

7.6 Active-Load Switching

Field-effect transistors, when operated in the ohmic region, can be used in place of resistors. On the macroscopic scale, it is more reasonable to use a resistor when a resistance is needed. For the manufacture of integrated circuit chips, resistors are avoided when a physically smaller FET can accomplish the same purpose. Example 7.5A is a simplified version of a MOSFET inverter. The top MOSFET has the drain wired to the gate of the transistor. This configuration places the MOSFET in the ohmic region where it serves as the drain resistance for the MOSFET below it.

Since the evaluation library contains power MOSFETs whose characteristics are different from those of the transistors in a CMOS integrated circuit chip, Example 7.5B is used for the simulation circuit. This means that the simulation does not accu-

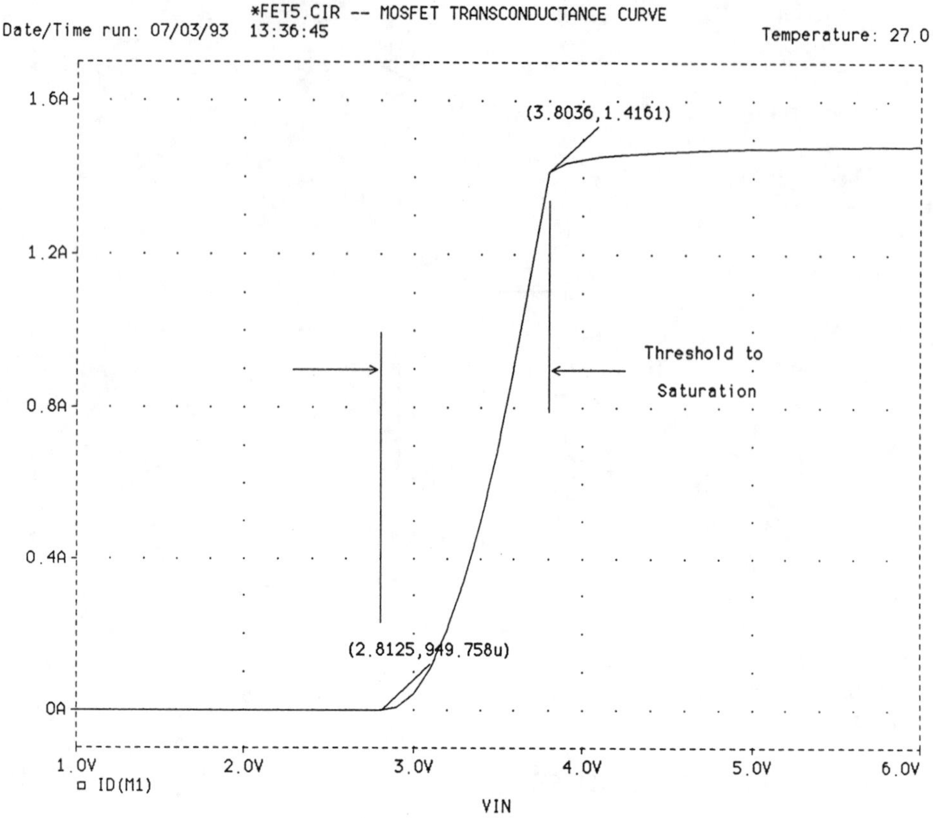

*FET5.CIR -- MOSFET TRANSCONDUCTANCE CURVE
Date/Time run: 07/03/93 13:36:45
Temperature: 27.0
1.6A
1.2A
0.8A
0.4A
0A
(3.8036,1.4161)
Threshold to
Saturation
(2.8125,949.758u)
1.0V
2.0V
3.0V
4.0V
5.0V
6.0V
ID(M1)
VIN

Figure 7.8

rately reflect the actual condition in a MOS inverter; but the principle is satisfactorily presented. The liberty taken is justified by considering that students seldom need to simulate lab experiments in making their own integrated circuit chips.

1
+5 V
2
3
IRF150
10 K
100 nF
0
A
1
+5 V
2
3
IRF150
10 K
100 nF
0
B

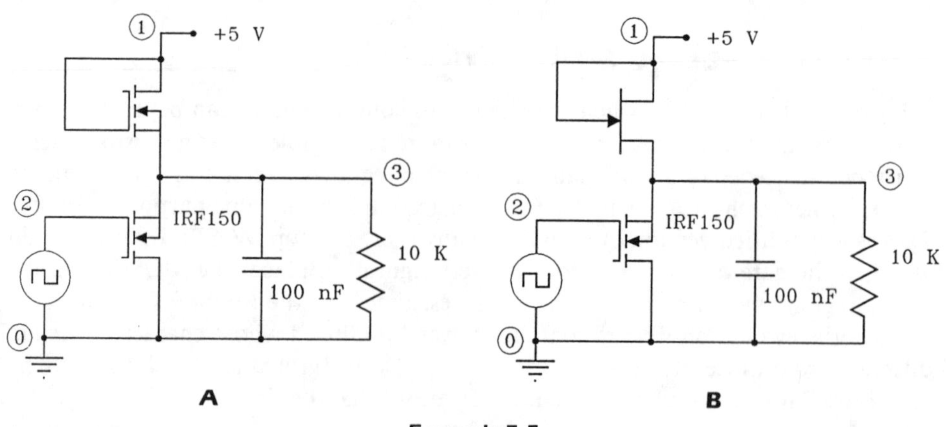

Example 7.5

```
*FET6.CIR—ACTIVE LOAD INVERTER
.INC \SPICE\FET.LIB
VSUPPLY 1 0 5
VIN 2 0 PULSE (0 5 0 1N 1N .5M 1M)
J1 1 1 3 J2N3819
M1 3 2 0 0 IRF150
C1 3 0 100N
RLOAD 3 0 10K
.TRAN .5M 5M
.PROBE
.END
```

The 100 nanofarad capacitor in parallel with the load resistor is not strictly necessary. It has been added to the simulation for filtering purposes. Without the capacitor, spikes are generated at the corners of the square wave output. The output rises to nearly 5 Volts, but it is .6 Volts above the expected low level. A low level closer to zero could be achieved by using a FET that was designed to provide an even higher drain resistance when the lower transistor was conducting. See Figure 7.9.

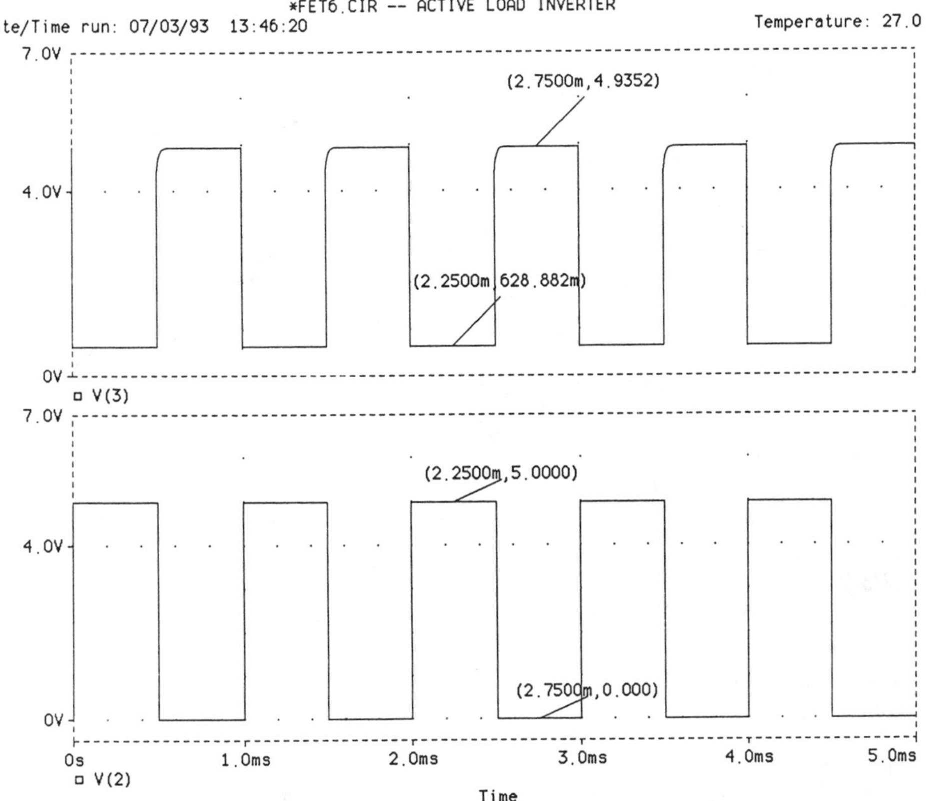

Figure 7.9

7.7 Cascode Amplifier

High-frequency applications call for amplifiers that have not only high input resistance but also high input capacitive reactance. The cascode amplifier configuration is one solution for providing high input reactance. In Example 7.6, a common-gate amplifier is driven by a common-source amplifier. The output shows a gain of 10 at 10 megaHertz, as shown in Figure 7.10.

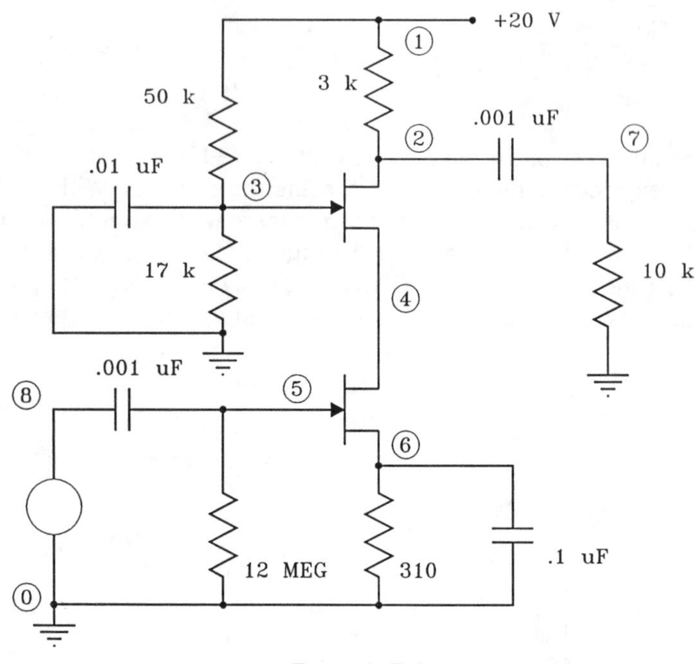

Example 7.6

```
*FET7.CIR–CASCODE AMPLIFIER
.INC \SPICE\FET.LIB
VSUPPLY 1 0 20
VIN 8 0 SIN( 0 .1 10MEG)
RDRAIN 1 2 3K
R1 1 3 50K
R2 3 0 17K
CGATE 3 0 10N
J1 2 3 4 J2N3819
J2 4 5 6 J2N3819
RSOURCE 6 0 310
CBYPASS 6 0 100N
CIN 8 5 1N
RBIAS 5 0 12MEG
```

```
COUT 2 7 1N
RLOAD 7 0 10K
.TRAN .1U .5U 0 1N
.PROBE
.END
```

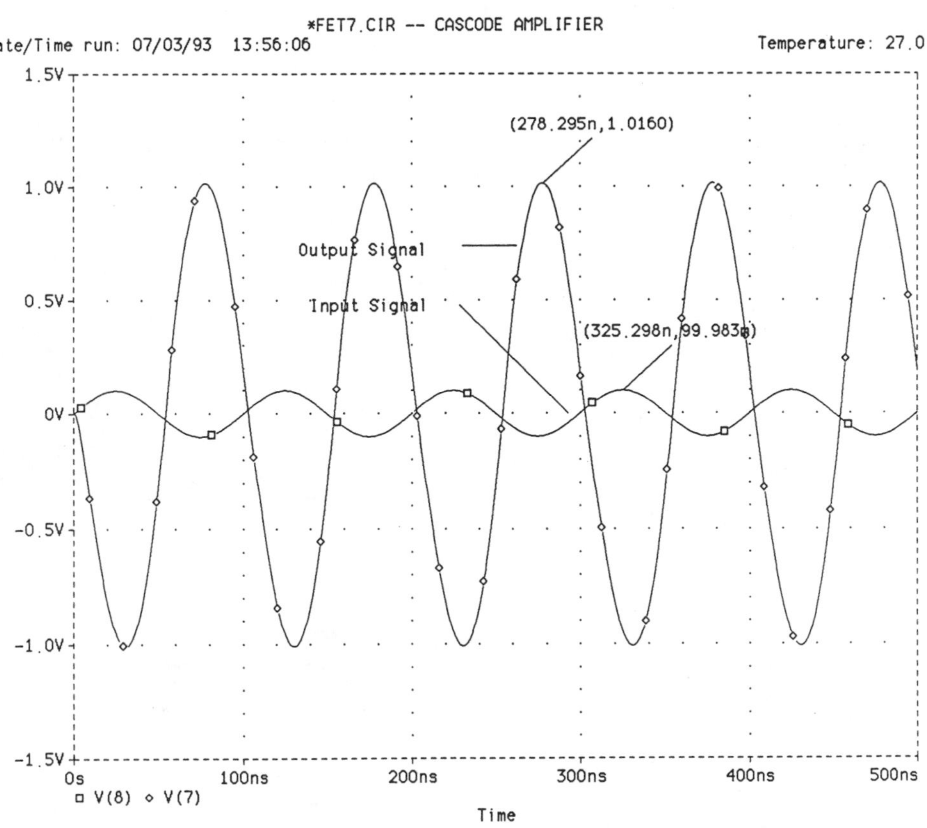

Figure 7.10

8

Analog Communications Circuits

The design and analysis of application circuits is the true purpose of circuit simulation software. Much time and effort can be saved if new designs are simulated and design parameters are adjusted using computers. It thus becomes unnecessary to build a physical prototype for the initial circuit and for each revision.

Up to this point PSpice is used strictly as a tool for educational purposes. Basic network theory is illustrated, along with the fundamentals of semiconductor theory. The simulations determine currents and voltage levels at interesting points in circuits, as well as the phase shifts caused by reactive elements. Amplification devices, such as op amps and transistors, are examined in terms of voltage gain and bandwidth characteristics.

As useful as simulations are for learning how components respond to applied voltages, the real power of PSpice is unlocked when it is used to design and test circuits intended to perform needed real-world functions. The field of analog communications provides the opportunity for the study of such circuits in most electronics curriculums. In this chapter, PSpice is used to design and examine oscillators, modulation and demodulation circuits, and frequency doublers—i.e., circuits typically studied in an analog communication class.

8.1 Colpitts Oscillator

In virtually all communications systems, the message signal is modulated onto a high-frequency carrier wave. The designer of the system can choose from many modulation schemes, but common to every method is the need to generate the carrier wave. Communications textbooks typically examine several types of transistor oscillators that produce high-frequency sinusoids. Hartley, Colpitts, Clapp, and Pierce oscillators are all straightforward and simple to build as lab exercises. The Colpitts oscillator was chosen at random for examination.

The central feature for oscillators, whether built from transistors or op amps, is the use of positive feedback, more specifically, the return of part of the output signal to the input of the device in such a way that the returned signal is exactly in phase with the original input. For the Colpitts oscillator shown in Example 8.1, positive feedback is achieved at the base of the transistor by two phase shifts of 180° each. The first shift is produced at the collector by the class A operation of the transistor. The second is the result of the ground point between the two equal capacitors in the output tank circuit. The frequency of oscillation is determined by the resonant frequency of the tank circuit.

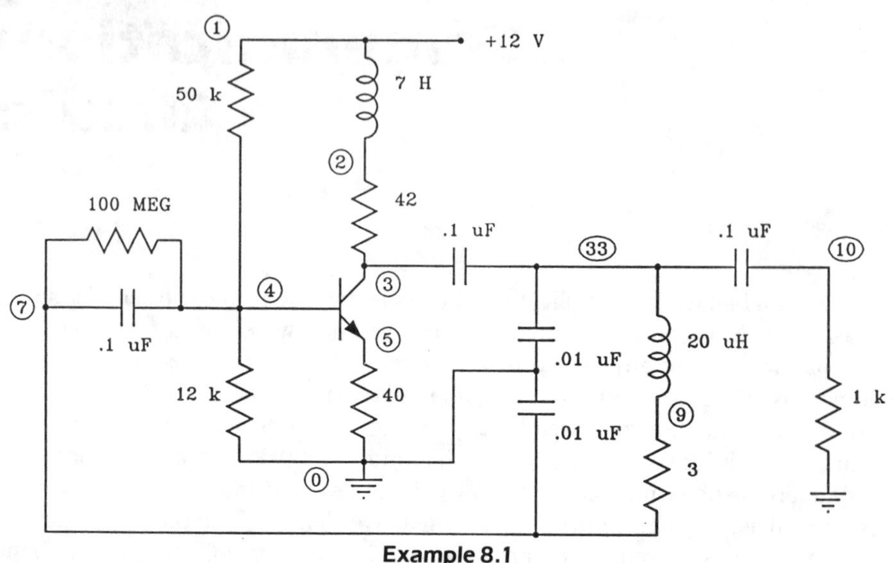

Example 8.1

```
*OSC.CIR–OSCILLATOR CIRCUIT, 500kHz
.INC \SPICE\TRAN.LIB
VSOURCE 1 0 12
LCHOKE 1 2 7
RCHOKE 2 3 42
R1BIAS 1 4 50K
R2BIAS 4 0 12K
Q1 3 4 5 Q2N2222A
REMITTER1 5 0 40
CIN 4 7 .1U
RBYPASS 4 7 100MEG
COUT 3 33 .1U
CTANK1 33 0 .010U
CTANK2 0 7 .010U
LTANK 33 9 20U
RTANK 9 7 3
CLOAD 33 10 .1U
RLOAD 10 0 1K
.TRAN 1U 150U 0 100n
.PROBE
.END
```

In this circuit, the total capacitance of the tank is .005 μF. This produces a resonant frequency of approximately 500 kHz. Several features of this simulation deserve special comment. First, since oscillators require a certain amount of time before the oscillation begins, reaches a maximum amplitude, and becomes stable, the PSpice analysis must be continued long enough to display the final stable waveform. This length of time, which is for all practical purposes indeterminable, is found by trial and error. As shown in Figure 8.1, for this oscillator, 150 microseconds was found to be adequate time for the oscillation to begin and become regular but not long enough to balance out the DC component. Because of the amount of memory and the length of time required for the simulation, it is not a good practice to make the analysis time excessively long.

The frequency of the simulated oscillator matches the frequency of the physical breadboarded circuit, at least within component tolerances. The amplitude of the simulated output may or may not compare well to experimental results. The amplitude of the output is very sensitive to losses in the output circuit. Changes of just two or three ohms in the inductor branch of the tank radically change the amplitude of the output signal. Since these losses can be quite difficult to measure in a breadboarded circuit, experimental and simulated amplitudes may differ considerably.

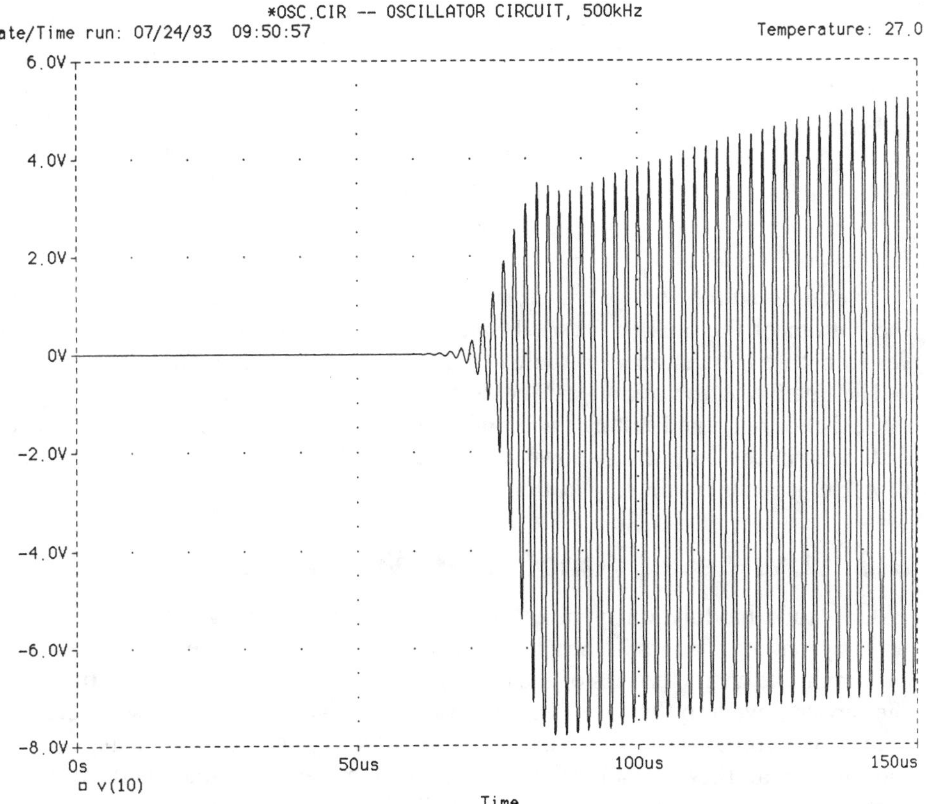

Figure 8.1

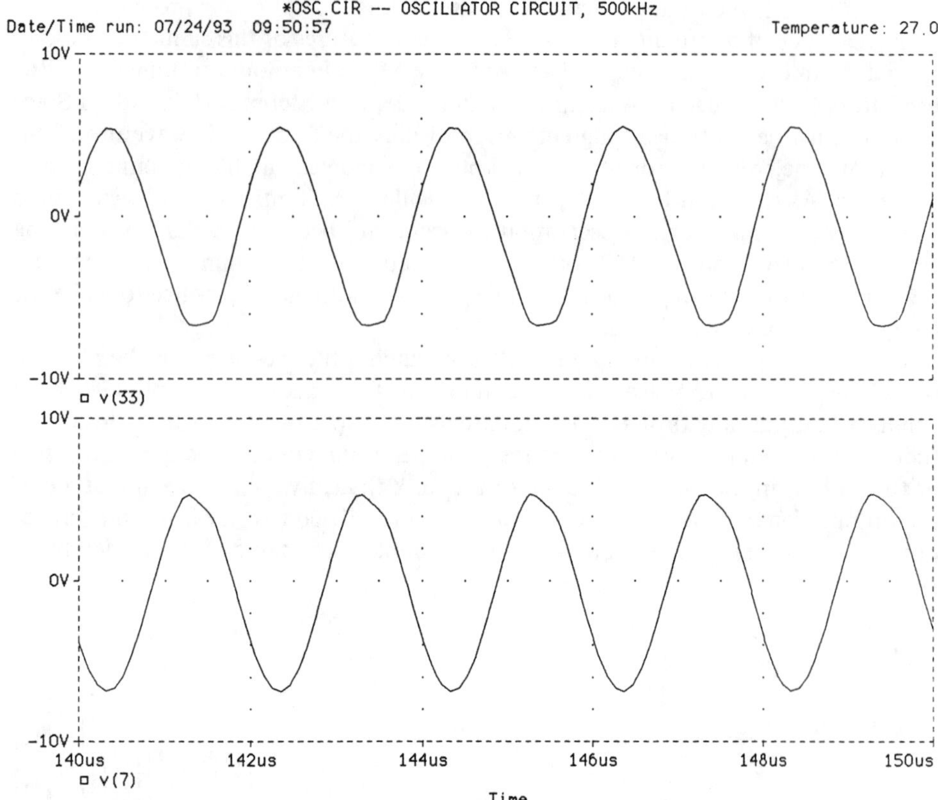

Figure 8.2

By examining the voltage waveform at the top of the tank circuit (point 33) and the bottom of the tank (point 7), PSpice can verify that the collector signal is inverted before it is returned to the base of the transistor. See Figure 8.2.

The quality of the output can be determined by performing a Fourier analysis (X-axis:Fourier) and noting the amplitude of the spurious frequencies that cause distortion in the output, as shown in Figure 8.3. Whether these distortions are acceptable depends on the application.

8.2 Low-Level Amplitude Modulation

Amplitude modulation involves impressing the message signal onto the carrier wave by making changes in the amplitude of the carrier to encode the frequency and the amplitude characteristics of the message signal. This can be achieved by feeding the carrier wave into the base of a class A amplifier and using the message signal to control the voltage at the emitter. As the emitter voltage changes in step with the message signal, the current through the transistor changes. Changes in the transistor current result in a varying gain for the radio frequency carrier at the base and an amplitude modulated signal.

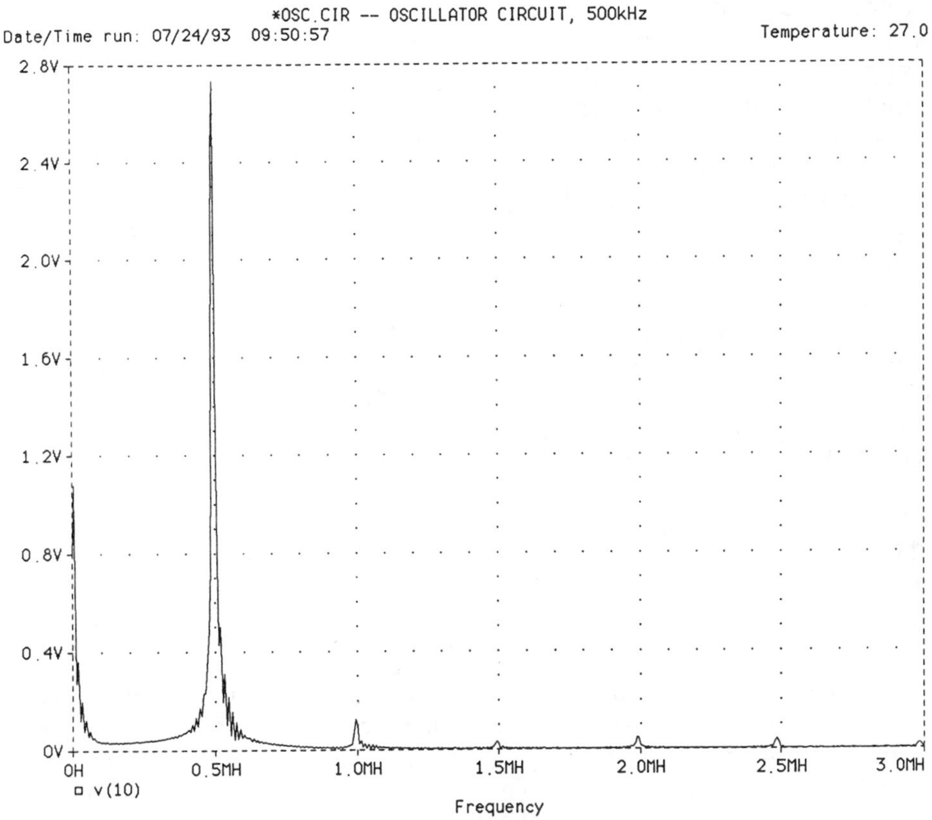

*OSC.CIR -- OSCILLATOR CIRCUIT, 500kHz
Date/Time run: 07/24/93 09:50:57 Temperature: 27.0

2.8V
2.4V
2.0V
1.6V
1.2V
0.8V
0.4V
0V
0H 0.5MH 1.0MH 1.5MH 2.0MH 2.5MH 3.0MH
□ v(10)
Frequency

Figure 8.3

This is called low-level modulation because the input signals and the resulting AM wave have small amplitudes. The modulating amplifier is usually followed by a linear amplifier to increase the amplitude before the signal is sent to the antenna. Example 8.2 is a typical modulator that can be found in most communications textbooks.

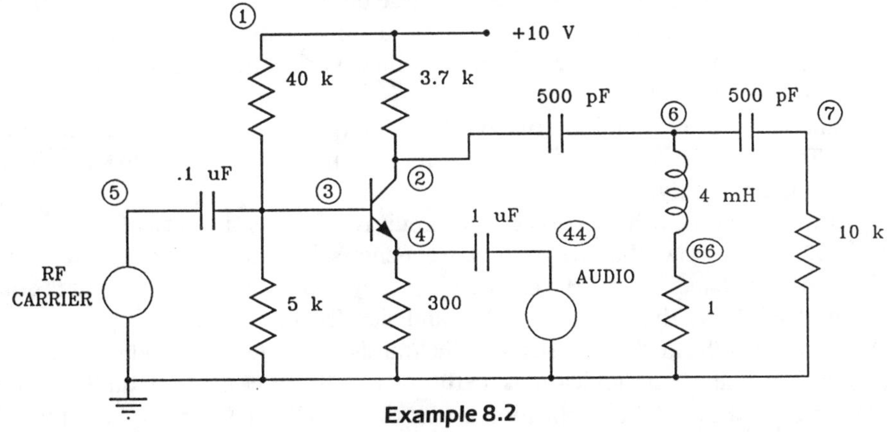

Example 8.2

```
*MOD.CIR—MODULATOR CIRCUIT
.INC \SPICE\TRAN.LIB
VSUPPLY 1 0 10
QMODULATOR 2 3 4 Q2N2222A
RCOLLECTOR 1 2 3.7K
REMITTER 4 0 300
RBIAS1 1 3 40K
RBIAS2 3 0 5K
CRF 5 3 .1U
VRF 5 0 SIN(0 .001 .1MEG)
CCOUPLE 4 44 1U
VINTELLIGENCE 44 0 SIN(0 .17 1K)
COUT1 2 6 500P
LFILTER 6 66 4M
RFILTER 66 0 1
COUT2 6 7 500p
RLOAD 7 0 10k
.TRAN .02M 2M
.PROBE
.END
```

The circuit description indicates a 1 kHz wave is used as the intelligence signal. In practice, the intelligence signal is typically music or a spoken message—that is, an aperiodic irregular signal. However, Fourier analysis theory proves that any signal, periodic or aperiodic, can be considered the sum of single-frequency sine waves. A circuit that performs well for each of the frequencies in that sum must also perform well for the combined waveform. A simultaneous mathematical analysis for all the frequencies in a voice signal would be, at the very least, difficult. Therefore a typical frequency is used.

The carrier frequency is a rather impractical 100 kHz. This has been done to speed the analysis and to avoid overtaxing the memory of the typical personal computer. To analyze the circuit for two cycles of the intelligence requires the analysis for 200 cycles of the carrier wave. This represents a large but acceptable amount of work for the computer. To raise the carrier frequency to a more realistic 1 MHz would be fine for the commercial version of PSpice running on a workstation, but is an unrealistic amount of work for a PC.

The analysis of the circuit is carried out for 2 milliseconds—two cycles of the intelligence. The resulting AM waveform, shown in Figure 8.4, indicates that the circuit is performing correctly.

A better indicator of the success of the circuit is a Fourier analysis of the output, Figure 8.5, which shows that there are no frequencies of significant amplitude except in the frequency band around the carrier frequency. Adjusting the X-axis, as is done in Figure 8.6, shows the expected carrier and side frequencies.

While injecting the intelligence signal at the transistor emitter is a standard technique found in almost any communications textbook, the "T" filter network on the output is not typically shown. The implication is that the signal taken directly from the col-

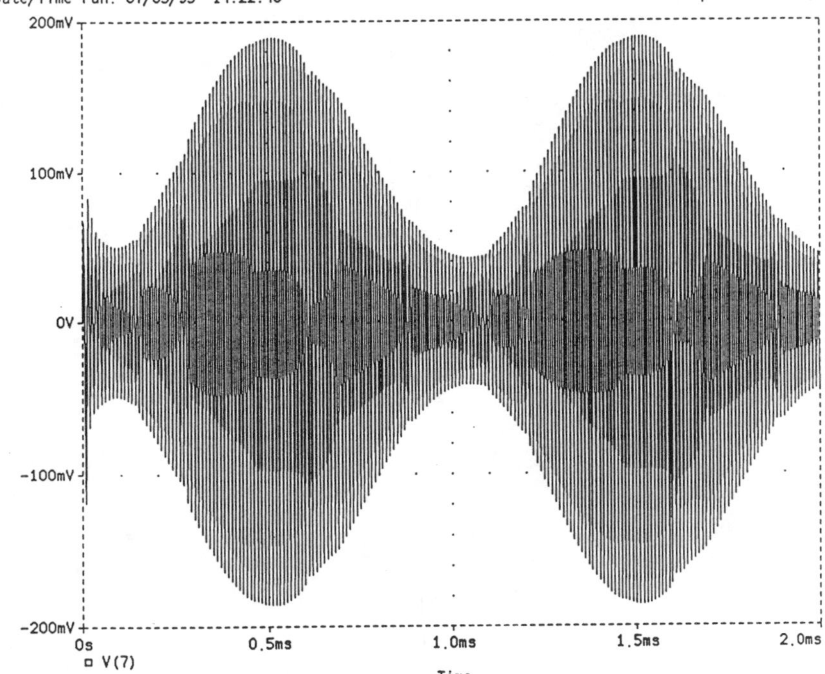

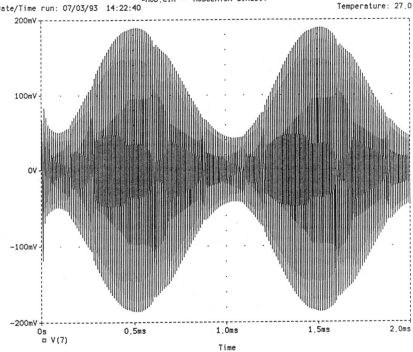

Figure 8.4

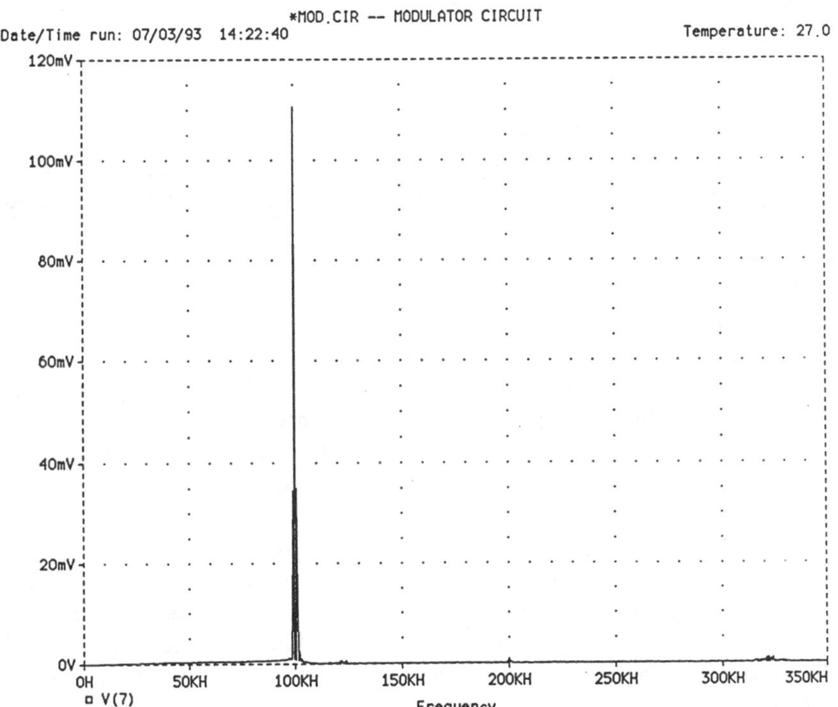

Figure 8.5

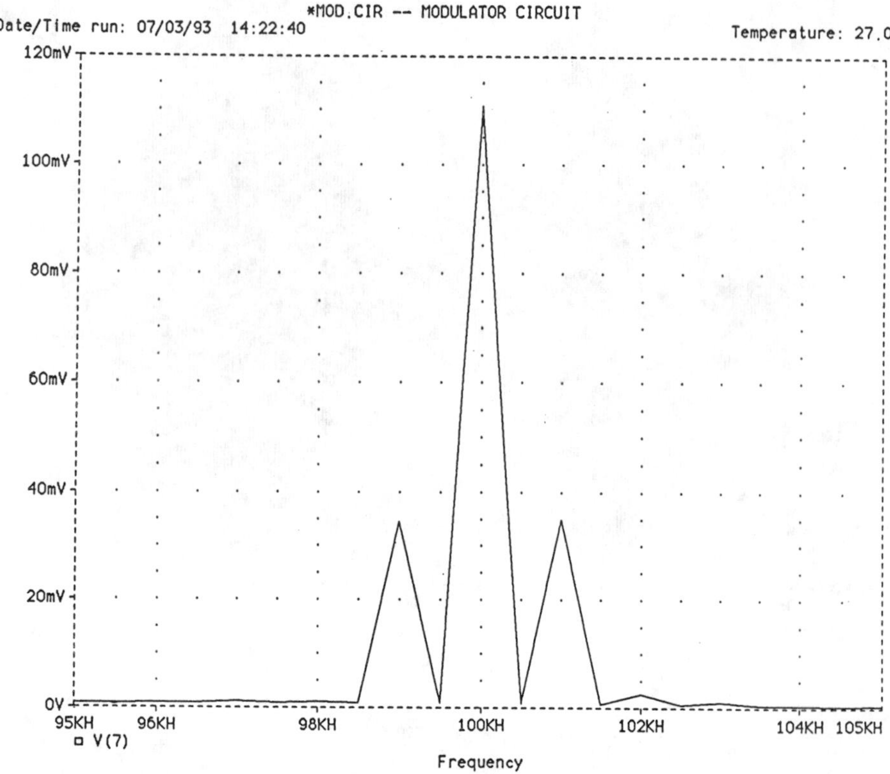

Figure 8.6

lector is an adequate AM signal. This can be examined by removing the filter, moving the load resistor to point 6 on the schematic, and running the analysis program.

```
COUT1          2 6 500P
RLOAD          6 0 10k
```

The plot shown in Figure 8.7 is the output at point 6.

Clearly the output from this modulator is unacceptable, but it does not indicate the nature of the problem or suggest how it can be solved. This is an opportunity to show the strength of simulation software as an aid in circuit design. By performing Fourier analysis on the unfiltered output at point 6, PSpice provides the information shown in Figure 8.8.

This plot indicates that the distortion in the output is due to the intelligence signal and its harmonics. This is information that would be available only to a designer who had access to a spectrum analyzer—a device usually too expensive for technical school budgets. The Fourier analysis also indicates that the solution to the problem is a high-pass filter that will leave the carrier and side frequencies but remove the unwanted low frequencies. In the original filtered design, the low frequencies are shorted to ground through the filter coil.

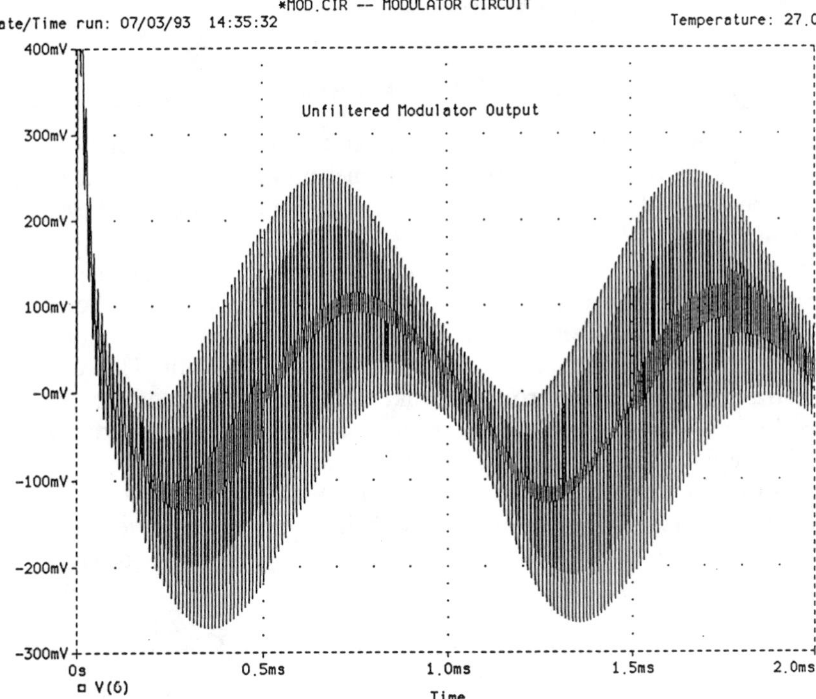

Figure 8.7

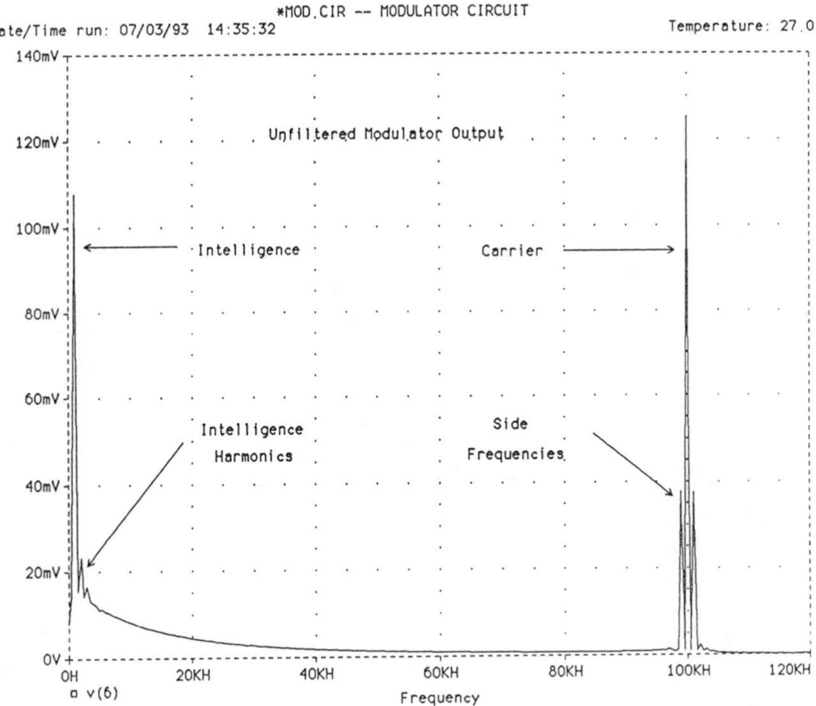

Figure 8.8

8.3 High-Level Amplitude Modulation

Because of the low efficiency of class A amplifiers, low-level modulation is usually confined to low-power transmitters. When higher power levels are required, there is a preference for the high efficiency of the class C amplifier and high-level modulation techniques.

In the design shown in Example 8.3, a 4 kHz intelligence is modulated onto a 455 kHz carrier. The design is not entirely typical of textbook designs for high-level modulation circuits. Typically the intelligence signal is added to the modulating amplifier by transformer coupling at point 2 above the tank circuit. The resulting AM signal is transformer coupled from the tank. Here, the use of transformers has been avoided because of the large amount of memory required to simulate transformer action. The audio enters the circuit through a capacitor that blocks the 12 Volt supply from shorting to ground, and the 7 Henry chock blocks the audio from shorting through the DC supply. The output of the circuit is taken to be the voltage generated across the coil in the tank circuit—i.e., the voltage from point 6 to point 7.

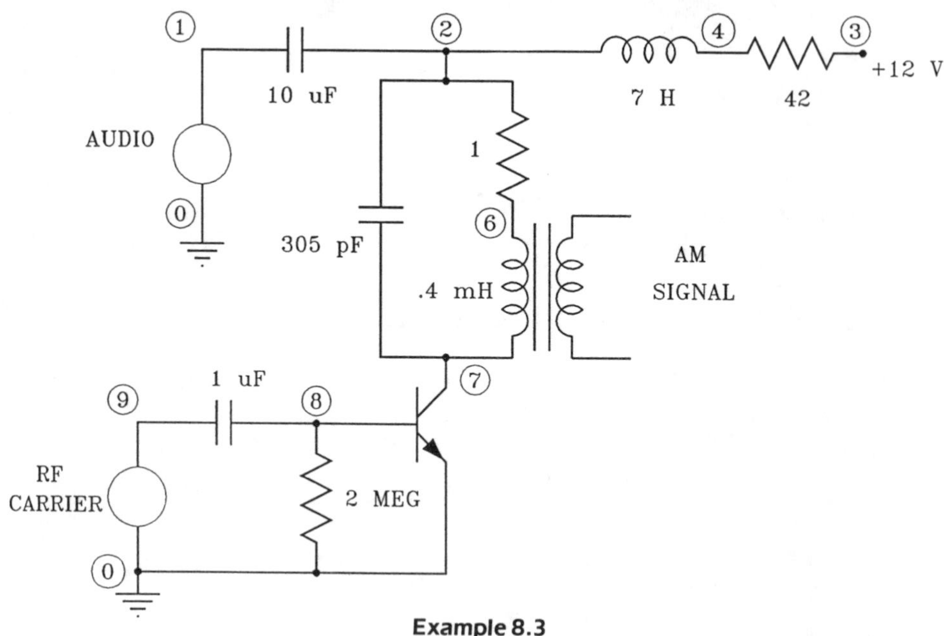

Example 8.3

```
*HMOD.CIR—HIGH LEVEL MODULATION
.INC \SPICE\TRAN.LIB
VAUDIO 1 0 SIN(0 6 4K)
CAUDIO 1 2 10U
VSUPPLY 3 0 12
RSUPPLY 3 4 42
LSUPPLY 4 2 7
```

```
CTANK 2 7 305P
RTANK 2 6 1
LTANK 6 7 .4M
QMODULATOR 7 8 0 Q2N2222A
RBIAS 8 0 2MEG
CRF 9 8 1U
VRF 9 0 SIN(0 1 455K)
.TRAN .05M .5M
.PROBE
.END
```

The resulting AM signal shows the inherent advantages of high-level modulation. First, the maximum peak-to-peak amplitude is approximately 100 times larger than the signal from the low-level modulator. Consequently the signal is ready to be sent to the antenna. Second, there is no need for subsequent filtering. The band pass characteristics of the tank circuit, shown in Figure 8.9, insure that only the carrier and the side frequencies are developed to significant amplitudes across the tank coil.

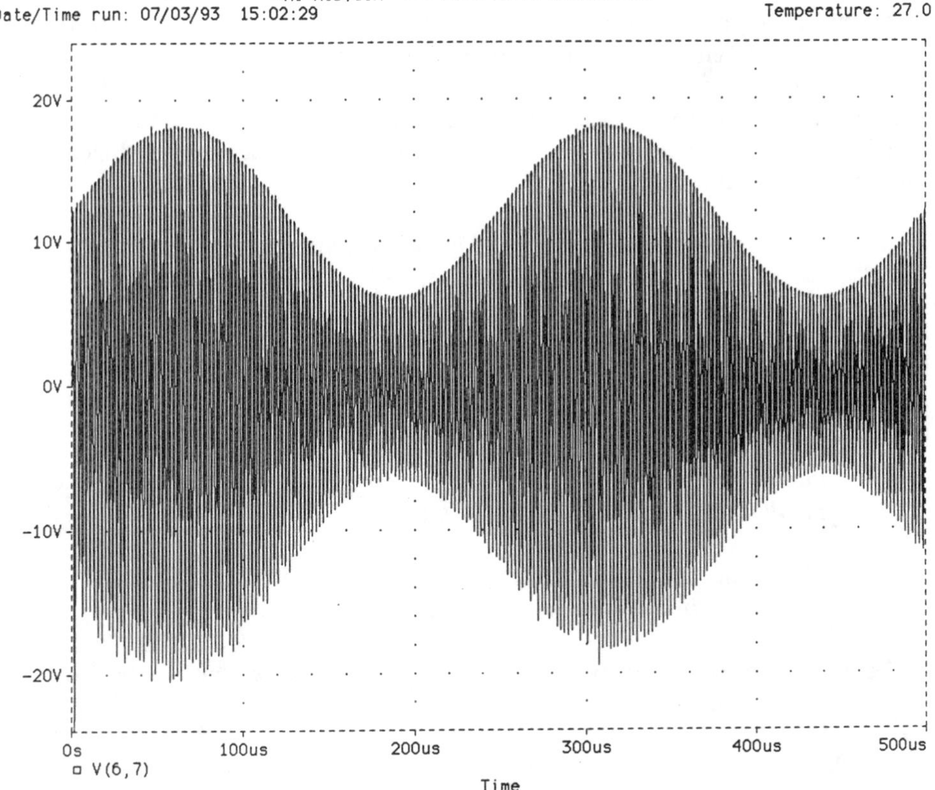

Figure 8.9

A visual inspection of the output waveform reveals no discernible distortion. Even though the analysis calculations have been minimized by avoiding the use of transformers in the circuit, there is inadequate memory to perform a Fourier analysis. The apparent lack of unwanted frequencies cannot be verified without a system that can access a larger amount of RAM memory.

8.4 AM and FM Demodulation

The demodulation of an AM waveform to recover the original intelligence signal can be accomplished with little more than a diode and a low-pass filter. However, before demodulation can take place there must be an amplitude-modulated signal to demodulate. The high-level modulator can provide an acceptable signal, but it uses so much memory that a demodulator cannot be added to the circuit. The low-level modulator uses less memory, but the signal must be amplified before a diode detector can be used. Adding a linear amplifier is not difficult, but once it is done, the user is again left with insufficient memory for the demodulation circuit.

The solution depends on the fact that one method for demodulating a frequency-modulated signal is to use a slope detector to produce an amplitude-modulated signal. This amplitude-modulated signal can then be passed through a diode detector and low-pass filter to recover the intelligence signal. This allows the demonstration of AM and FM demodulation in a single circuit.

A single frequency, frequency-modulated (SFFM) waveform is included as a standard input signal in the PSpice software. This voltage source allows the designer to specify the parameters of an FM signal modulated by a specified intelligence frequency. The general form is

V<name> <pos node> <neg node> SFFM (wave parameters)

where the wave parameters are defined as follows:

SFFM (offset voltage
peak amplitude
carrier frequency
modulation index
intelligence frequency)

Offset Voltage This value is usually set to zero, but it gives the user the opportunity to add a DC offset voltage.

Peak Amplitude FM waveforms are characterized by a constant peak voltage specified in the second parameter.

Carrier Frequency Center or rest frequency for the waveform.

Modulation Index This is the ratio of the deviation from the center frequency to the intelligence frequency that causes the deviation.

Intelligence Frequency This is the frequency of the waveform to be modulated onto the carrier.

With this readymade FM signal, Example 8.4 suffices for a simple demodulation circuit.

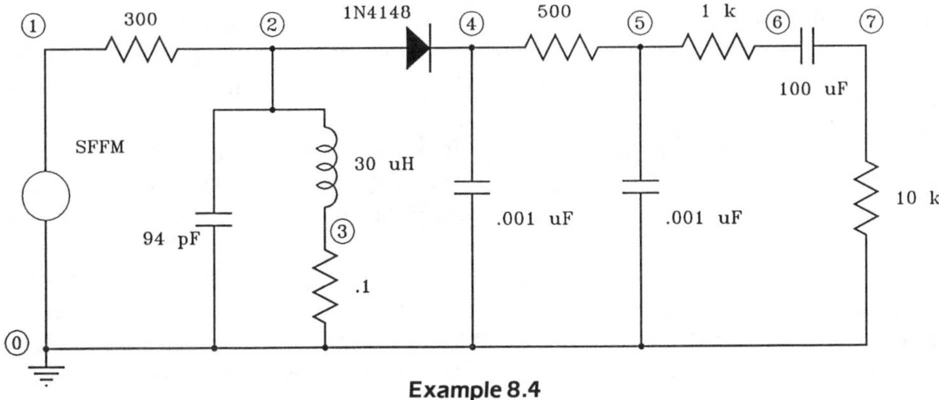

Example 8.4

```
*FM.CIR—FM DEMODULATOR
.INC \SPICE\DIODE.LIB
VFM        1 0 SFFM(0 2 1MEG 80 5K)
R1         1 2 300
CTANK      2 0 94P
LTANK      2 3 30U
RTANK      3 0 .1
D1         2 4 D1N4148
CFILTER        4 0 1N
RFILTER        4 5 500
CFILTER2       5 0 1N
RFILTER2       5 6 1000
COUT           6 7 100U
RLOAD          7 0 10K
.TRAN      40U 400U
.PROBE
.END
```

The circuit description indicates that the input wave is a 1 MHz carrier modulated with a 5 kHz intelligence signal. The amplitude is 2 volts and the modulation index is 80. A modulation index of 80 is, in fact, unrealistically large for a 5 kHz signal. The 5 kHz frequency and 80 modulation index were chosen to make the features to be examined more prominent without exceeding memory limitations. In commercially produced FM receivers, more sophisticated methods for FM demodulation are used that do not require the input to have such a high modulation index. The single slope method used in the above circuit is the simplest and least effective method for FM demodulation.

The single-slope detector uses a tank circuit with a resonant frequency that is somewhat higher than the carrier frequency of the FM signal. In this circuit, the resonant frequency is 3 MHz. The amount of deviation in an FM signal is proportional to the amplitude of the intelligence—i.e., a loud message signal causes a greater swing from the rest frequency than a quiet signal. The band-pass circuit passes a large signal when the FM signal frequency is nearer to the resonant frequency and a smaller amplitude when the signal frequency is low and farther from resonance. In short, the FM signal is converted to an AM signal. The signal at point 2 is shown in Figure 8.10.

This points out another advantage to software simulation. This signal cannot be viewed on an oscilloscope unless the scope is a digital storage scope. Ordinary oscilloscopes are unable to properly trigger on an FM signal because of the rapidly changing signal frequency.

Closer inspection of the signal at point 2 would reveal that the frequency modulation characteristics of the waveform have not been lost. The tank circuit merely gives low frequencies a small peak-to-peak amplitude and high frequencies a large peak-to-peak amplitude. Once this is observed, it should be noted that this fact is quite ir-

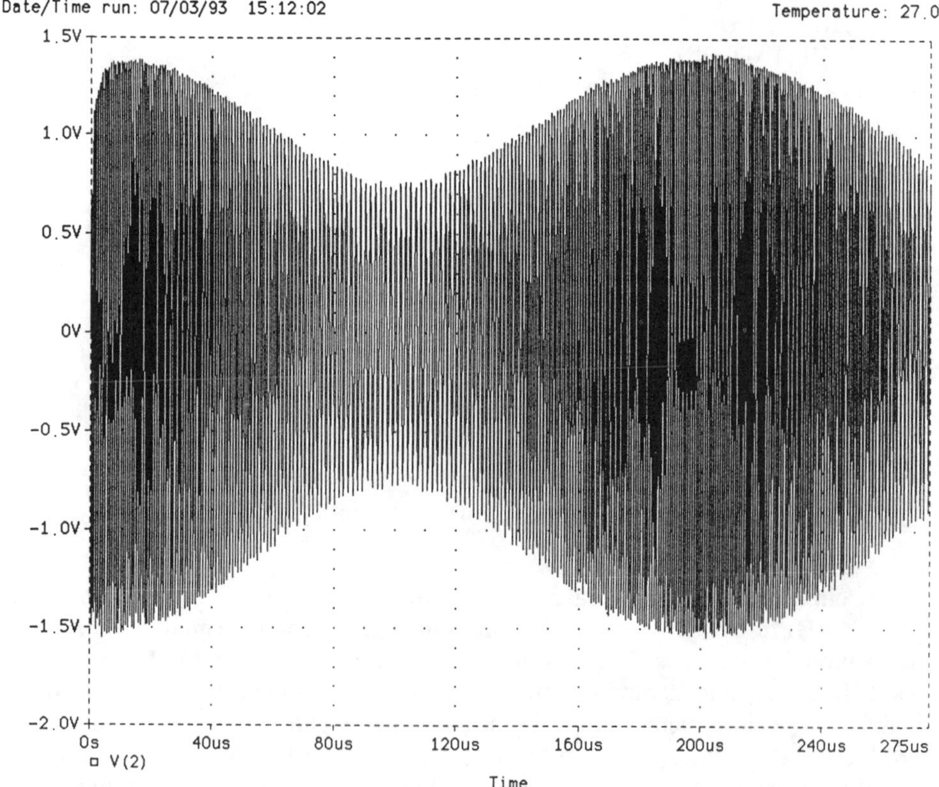

Figure 8.10

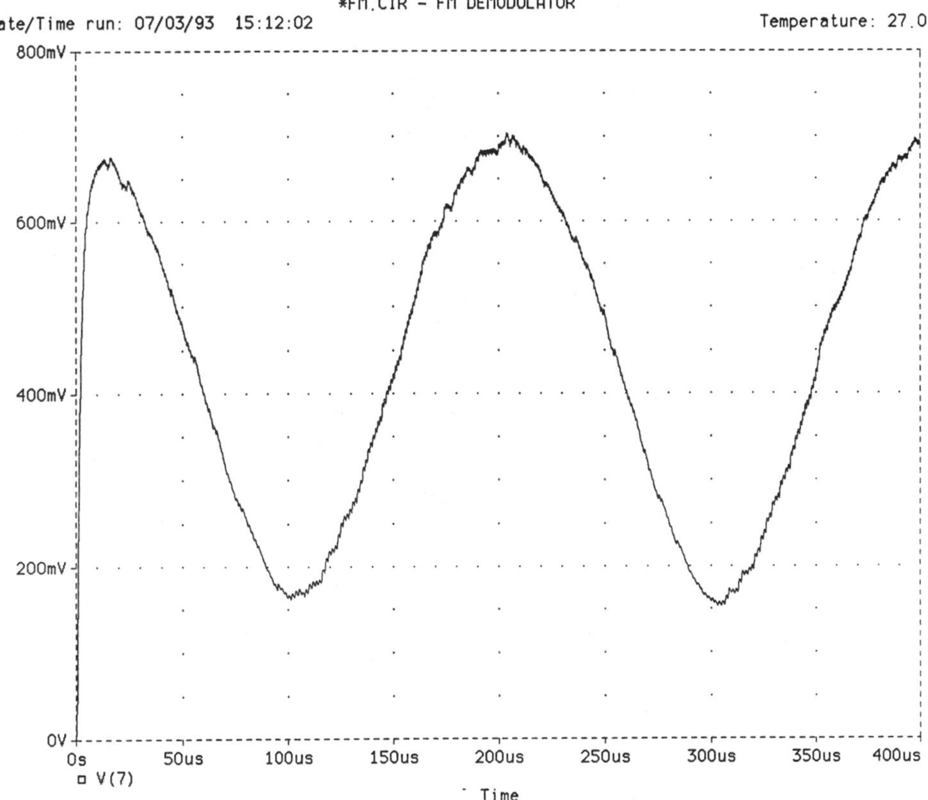

Figure 8.11

relevant since the diode and following low-pass filter will create a new signal that traces the envelope of the AM signal. A two-stage filter was used to produce a cleaner signal at the load as shown in Figure 8.11.

The remaining question is to ask how the 3 MHz resonant frequency for the tank circuit was determined? With enough time and thought, the value could be calculated, but it is far simpler and more time saving to merely guess at some value for resonance, model the band-pass circuit, and make necessary adjustments. The circuit shown in Example 8.5 is essentially the first half of the FM demodulator, except that the FM signal has been replaced by a sine wave set to sweep the frequencies from 1 kHz to 10 GHz. Figure 8.12 graphs the frequency response at point 2 and shows that a 1 MHz input falls on the side of the band-pass curve where the frequency response is somewhat linear.

```
*FM.CIR—FM DEMODULATOR
VFM      1 0 AC 2
R1       1 2 300
CTANK    2 0 94P
```

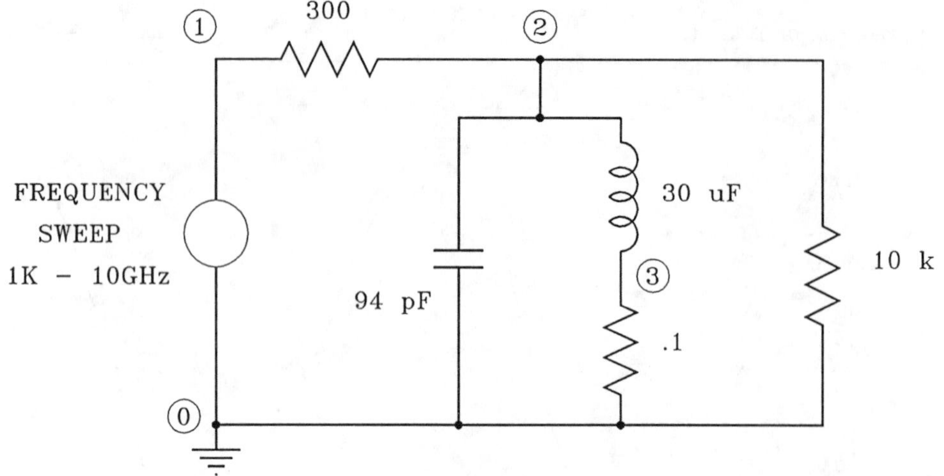

Example 8.5

Figure 8.12

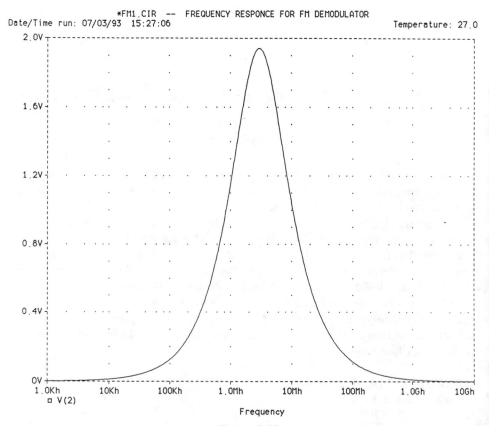

```
LTANK      2 3 30U
RTANK      3 0 .1
RLOAD          2 0 10K
.AC DEC   100 1000 10G
.PROBE
.END
```

For the purpose of converting FM to AM, the ideal condition would be to have a filter response that is perfectly linear in the frequency range of the FM signal. Because there is no linear range on a tank circuit frequency response, it is necessary to settle for a relatively linear range. To see the linearity of the curve, it is necessary to change the scaling of the X-axis from logarithmic to linear (X-axis:Linear) and to reduce the range to 100 kHz to 10 MHz. The resulting plot, Figure 8.13, indicates that for the signal bandwidth (approximately 1 MHz ± 170 kHz) the frequency response is nearly linear.

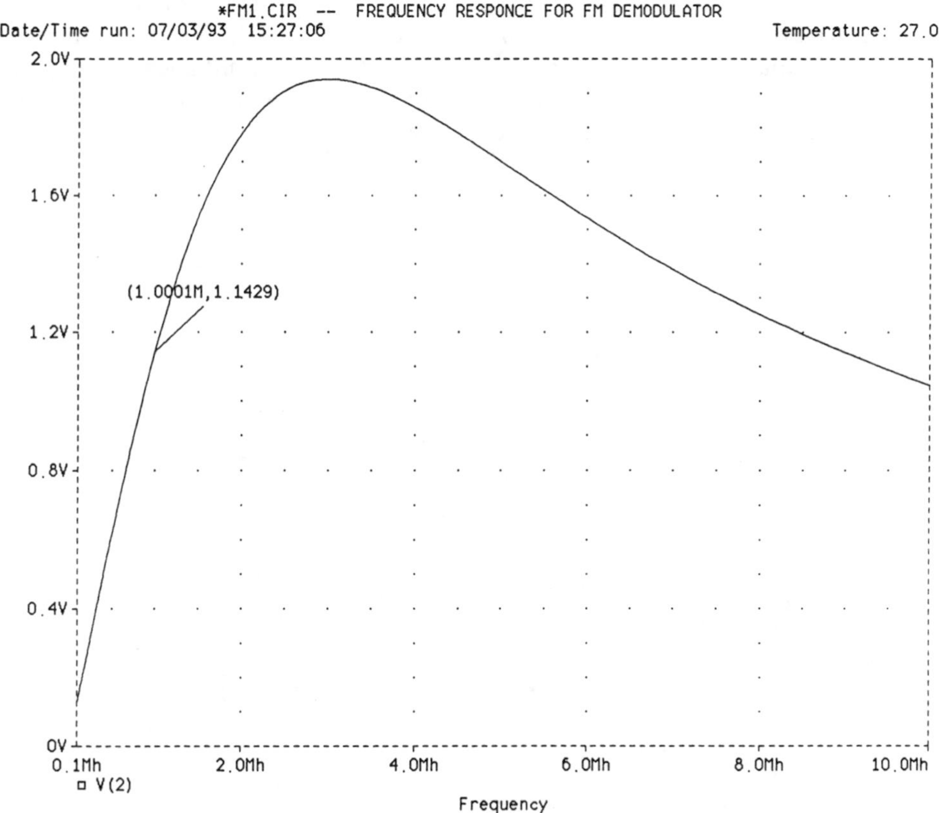

Figure 8.13

8.5 Doubler Circuits

The generation of a commercial FM signal requires that an extremely stable carrier be pulled from its rest frequency by the intelligence signal. Carrier stability means that the oscillator generating the carrier can have very little frequency drift during operation. The Colpitts oscillator examined in this chapter has too much drift for most applications because the component values in the tank circuit vary significantly with ambient temperature changes. Greater stability can be achieved with a crystal oscillator. However, crystal oscillators are so extremely stable that it is difficult to modulate a crystal to any significant deviation.

One solution to this dilemma is the Crosby FM transmitter. This design involves modulating a relatively low-frequency carrier, using frequency multipliers to raise the modulated signal to the final transmission frequency, and then comparing the signal to a crystal reference frequency. Any difference between the carrier of the modulated signal and the reference is returned to the original oscillator, and the oscillator is automatically adjusted to compensate for the drift. While the entire Crosby transmitter is too complex to model with the evaluation version of PSpice, the frequency multiplier is worth examining.

Frequency multipliers consist of frequency doublers, triplers, or quadruplers, either singularly or in cascade. The principle for each is the same. A signal is injected into a device that produces a large amount of harmonic distortion. Then a band-pass filter is used to remove all unwanted signals. In Example 8.6, a 1 MHz sine wave is given to a class C amplifier. The tank circuit in the collector branch has a resonant

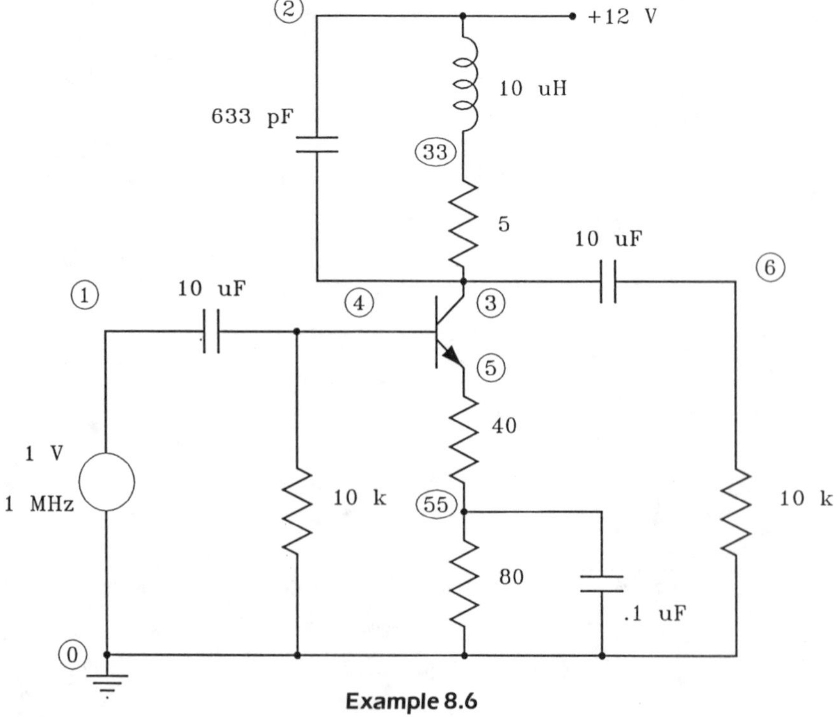

Example 8.6

frequency of 2 MHz. The class C amplifier produces and amplifies many harmonics of the input signal, but all except the second harmonic are filtered out by the collector tank circuit.

```
*DOUBLER CIRCUIT
.INC \SPICE\TRAN.LIB
VSIGNAL 1 0 SIN( 0 1 1MEG)
VSUPPLY 2 0 12
CIN 1 4 10U
RBIAS 4 0 10K
LTANK 2 33 10U
CTANK 2 3 633P
RTANK 33 3 5
REMITTER 5 55 40
REMITTER2 55 0 80
CBYPASS 55 0 .1U
Q1 3 4 5 Q2N2222A
COUT 3 6 10U
RLOAD 6 0 10K
.TRAN .5U 20U 10U .1U
.PROBE
.END
```

In Figure 8.14, the input and output signals have been plotted on separate graphs by using "Plot_control" from the main PROBE menu.

Although the time-based graphs indicate the input signal has been doubled, the Fourier analysis shown in Figure 8.15 yields more complete information. Here it can be seen that while the second harmonic is the largest component of the output waveform, the original input and the third harmonic are of significant amplitude. Additional filtering can be accomplished by buffering the output with a FET transistor configured as a source follower and using a second tank circuit, as shown in Example 8.7.

```
*DOUBLER CIRCUIT
.INC \SPICE\TRAN.LIB
.INC \SPICE\FET.LIB
VSIGNAL 1 0 SIN( 0 1 1MEG)
VSUPPLY 2 0 12
CIN 1 4 10U
RBIAS 4 0 10K
LTANK 2 33 10U
CTANK 2 3 633P
RTANK 33 3 5
REMITTER 5 55 40
REMITTER2 55 0 80
CBYPASS 55 0 .1U
```

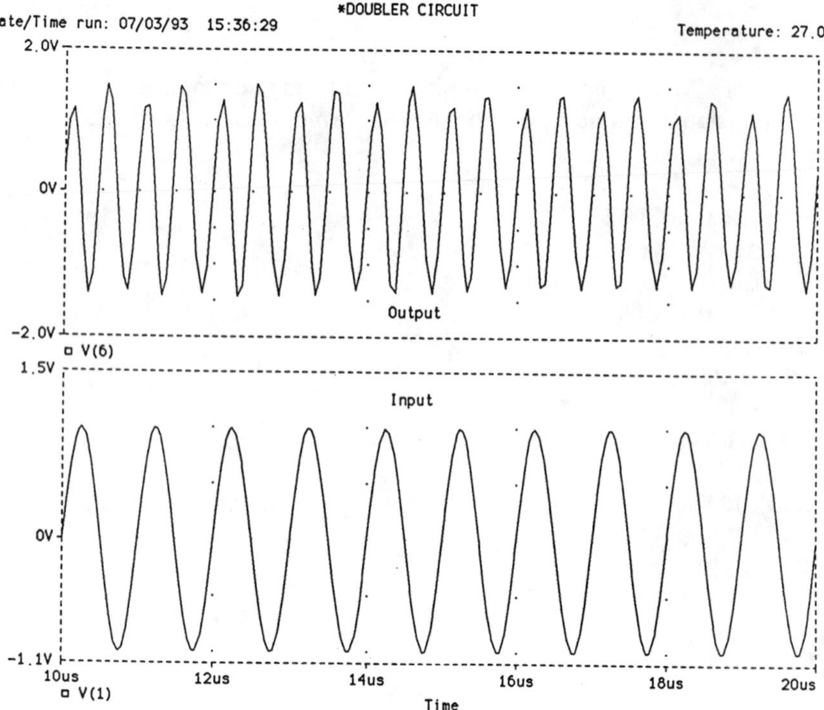

Figure 8.14

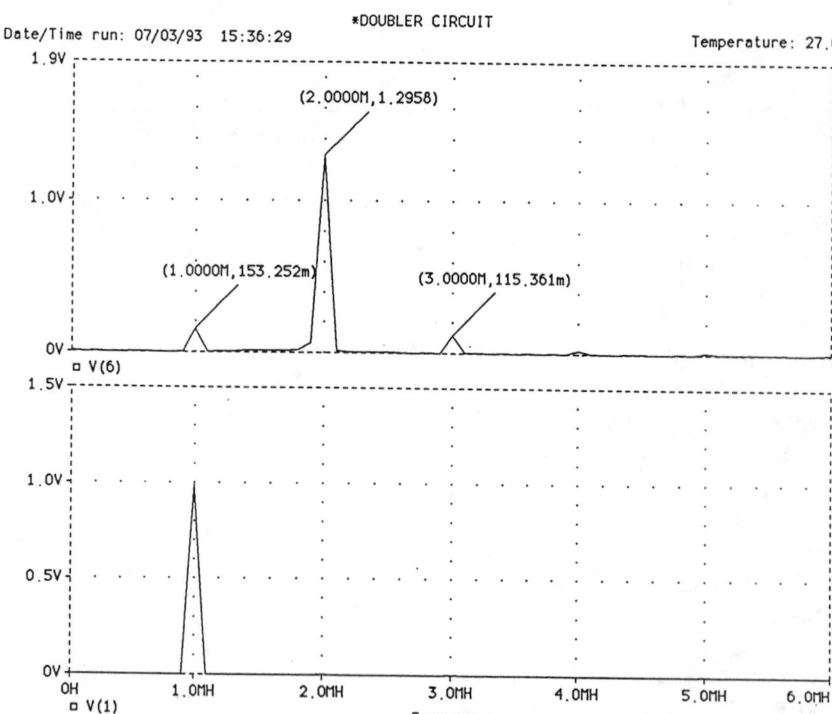

Figure 8.15

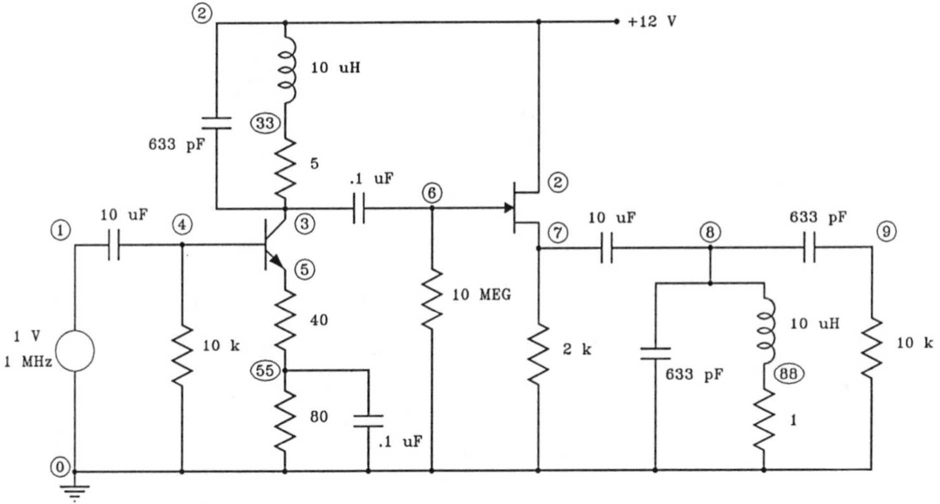

Example 8.7

```
Q1 3 4 5 Q2N2222A
COUT 3 6 10U
J1 2 6 7 J2N3819
RBIAS2 6 0 10MEG
RSOURCE 7 0 2K
COUT2 7 8 10U
CTANK2 8 0 633P
LTANK2 8 88 10U
RTANK2 88 0 1
LOUT3 8 9 10U
COUT3 9 10 633P
RLOAD 10 0 10K
.TRAN .5U 20U 10U .1U
.PROBE
.END
```

A Fourier analysis of the output voltage, Figure 8.16, now shows that the fundamental and the third harmonic have been reduced by a factor of approximately three.

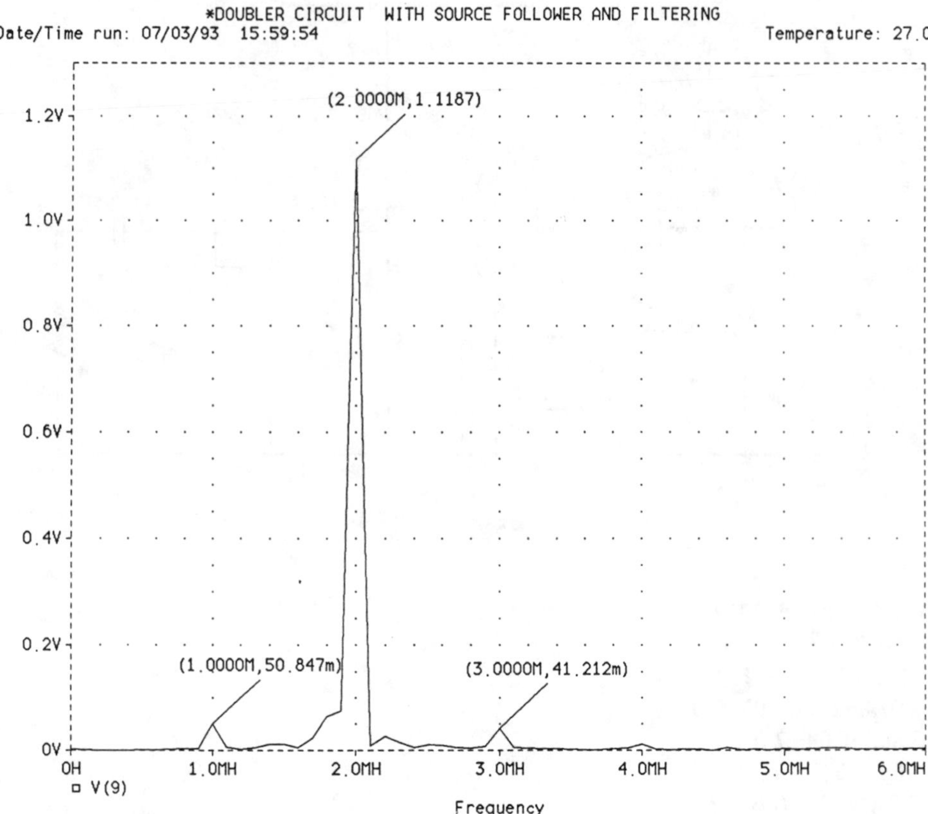

*DOUBLER CIRCUIT WITH SOURCE FOLLOWER AND FILTERING
Date/Time run: 07/03/93 15:59:54
Temperature: 27.0
(2.0000M,1.1187)
1.2V
1.0V
0.8V
0.6V
0.4V
0.2V
0V
(1.0000M,50.847m)
(3.0000M,41.212m)
0H
1.0MH
2.0MH
3.0MH
4.0MH
5.0MH
6.0MH
V(9)
Frequency

Figure 8.16

Appendixes

A Numeric Value and Expression Conventions

B Abridged PSpice Command List

C Abridged List of Analog Devices and Libraries

Numeric Value and Expression Conventions

Literal numeric values are written in standard floating point notation. PSpice assumes default units for numbers describing component values and electrical quantities. However, values can be scaled by following the number by the appropriate scale suffix as shown in Figure A-1.

Figure A-1: Scale Suffixes for Numeric Values

Symbol	Scale	Name
F	10^{-15}	femto-
P	10^{-12}	pico-
N	10^{-9}	nano-
U	10^{-6}	micro-
MIL	$25.4*10^{-6}$	--
M	10^{-3}	milli-
K	10^{+3}	kilo-
MEG	10^{+6}	mega-
G	10^{+9}	giga-
T	10^{+12}	tera-
C		Clock cycle†

†Clock cycle will vary and must be set where applicable.

Numeric values can also be indirectly represented by parameters (see the .PARAM command in Appendix B). Numeric values and parameters can be used together to form arithmetic expressions. PSpice expressions may incorporate the intrinsic functions shown in Figure A-2.

Figure A-2: Valid Functions for PSpice Expressions

Function	Meaning	Comments
ABS(x)	$\|x\|$	
SQRT(x)	$x^{1/2}$	
EXP(x)	e^x	
LOG(x)	ln(x)	log base e
LOG10(x)	log(x)	log base 10
PWR(x,y)	$\|x\|^y$	
PWRS(x,y)	$+\|x\|^y$ (if x>0), $-\|x\|^y$ (if x<0)	
SIN(x)	sin(x)	x in radians
COS(x)	cos(x)	x in radians
TAN(x)	tan(x)	x in radians
ATAN(x)	$\tan^{-1}(x)$	result in radians
ARCTAN(x)	$\tan^{-1}(x)$	result in radians
TABLE($x,x_1,y_1,x_2,y_2,\ldots x_n,y_n$)		result is the y value corresponding to x, when all of the x_n,y_n points are plotted and connected by straight lines. If x is greater than the max x_n, then the value is the y_n associated with the largest x_n. If x is less than the smallest x_n, then the value is the y_n associated with the smallest x_n.
LIMIT(x,min,max)		result is min if x < min, max if x > max, and x otherwise.

Most numeric specifications in PSpice allow arithmetic expressions. Some exceptions do exist and are summarized in the "Circuit File Construction" chapter in the *Circuit Analysis User's Guide.* There are also some differences between the intrinsic functions available in PSpice and those available in Probe. The "Probe" chapter in this manual presents the available intrinsic functions. A summary of the differences between PSpice and Probe functions can be found in the "Waveform Analysis" chapter in the *Circuit Analysis User's Guide.*

Abridged PSpice Command List

Figure B-1: Command Summary

Type	Command	Description
Standard Analyses	.DC .OP .TF .SENS .AC .NOISE .TRAN .FOUR	DC sweep Bias point Small-signal transfer DC sensitivity Frequency response Noise Transient Fourier components
Simple Multi-Run Analyses	.STEP .TEMP	Parametric Temperature
Statistical Analyses	.MC .WCASE	Monte Carlo Sensitivity/Worst-Case
Initial Conditions	.IC .NODESET .SAVEBIAS .LOADBIAS	Clamped bias point calculation Unclamped bias point calculation Stored .NODESET bias point Restored .NODESET bias point
Device Modeling	.MODEL .SUBCKT .ENDS .DISTRIBUTION	Modeled device definition Start subcircuit definition End subcircuit definition Model parameter tolerance distribution
Output Control	.PLOT .PRINT .PROBE .WATCH .WIDTH	Analysis plot to output file Analysis table to output file Simulation results to Probe data file View simulation results in progress Character width of output file

Figure B-1: Command Summary (Continued)

Type	Command	Description
Circuit File Processing	.FUNC .PARAM .END .INC .LIB	Expression function definition Parameter definition End of circuit simulation description Include specified file Reference specified library
Miscellaneous	.OPTIONS .TEXT	Sets miscellaneous simulation limits, analyses control parameters, and output characters Text expression, parameter, or file name used by digital devices

.AC — AC Analysis

General Form

.AC *<sweep type>* *<points value>*
+ *<start frequency value>* *<end frequency value>*

Examples

```
.AC LIN 101 100Hz 200Hz
.AC OCT 10 1kHz 16kHz
.AC DEC 20 1MEG 100MEG
```

The .AC statement is used to calculate the frequency response of a circuit over a range of frequencies. *<sweep type>* must be either LIN, OCT, or DEC, and *<points value>* is the number of points in the sweep.

LIN — Linear sweep. The frequency is swept linearly from the starting to the ending frequency. *<points value>* is the total number of points in the sweep.

OCT — Sweep by octaves. The frequency is swept logarithmically by octaves. *<points value>* is the number of points per octave.

DEC — Sweep by decades. The frequency is swept logarithmically by decades. *<points value>* is the number of points per decade.

Exactly one of LIN, OCT, or DEC, must be specified.

<end frequency value> **must not be less than** *<start frequency value>*, and both must be greater than zero. The whole sweep may specify only one point if you wish.

The frequency response is calculated by linearizing the circuit around the bias point. All independent voltage and current sources which have AC values are inputs to the circuit.

.PRINT, .PLOT, or .PROBE statements must be used to get the results of the AC sweep analysis.

If you specify group delay ("G" suffix) as an output, you must be sure that the frequency steps are close enough together that the phase of that output changes smoothly from one frequency to the next. Group delay is calculated by subtracting the phases of successive outputs and dividing by the frequency increment.

During AC analysis, the only independent sources which have non-zero amplitudes, are those with AC specifications. The SIN specification does not count as it is used only during transient analysis. See the "Specifying Analog Devices; Independent Sources" section of the "Analog and Digital Parts" chapter in the *Circuit Analysis User's Guide* for more information.

AC analysis is a linear analysis. To analyze non-linear functions, such as mixers, frequency doublers, AGC, etc. it is necessary to use transient analysis. If using Analog Behavioral Modeling, see the "Analog Behavioral Modeling" chapter in the *Circuit Analysis User's Guide.*

.DC DC Analysis

General Forms

.DC *<linear sweep type> <sweep variable name>*
+ *<start value> <end value> <increment value>*
+ [*nested sweep specification*]

.DC *<logarithmic sweep type> <sweep variable name>*
+ *<start value> <end value> <points value>*
+ [*nested sweep specification*]

.DC *<sweep variable name>* LIST *<value>**
+[*nested sweep specification*]

Examples

```
.DC VIN -.25 .25 .05
.DC LIN I2 5mA -2mA 0.1mA
.DC VCE 0V 10V .5V IB 0mA 1mA 50uA
.DC RES RMOD(R) 0.9 1.1 .001
.DC DEC NPN QFAST(IS) 1E-18 1E-14 5
.DC TEMP LIST 0 20 27 50 80 100
.DC PARAM Vsupply 7.5 15 .5
```

The .DC statement causes a DC sweep analysis to be performed on the circuit. The DC sweep analysis calculates the circuit's bias point over a range of values for *<sweep variable name>*. The first form, and the first four examples, are for doing a linear sweep. The second form, and the fifth example, are for doing a logarithmic sweep. The third form, and the sixth example, are for using a list of values for the sweep variable.

For linear sweeps,*<start value>* may be greater or less than *<end value>*: that is, the sweep may go in either direction. *<increment value>* must be greater than zero. For logarithmic sweeps (DEC or OCT), *<start value>* must be positive and less than *<end value>*. *<points value>* must be greater than zero.

A nested sweep is available (see third example). A second sweep variable, sweep type, start, end, and increment values may be placed after the first sweep. In this case the first sweep will be the "inner" loop: the entire first sweep will be done for each value of the second sweep. The rules for the values in the second sweep are the same as for the first. The second sweep generates an entire .PRINT table or .PLOT plot for each value of the sweep. Probe allows nested sweeps to be displayed as a family of curves.

The sweep can be linear, logarithmic, or a list of values. For [*linear sweep type*], the keyword LIN is optional, but either OCT or DEC must be specified for the *<logarithmic sweep type>*. The sweep types are:

LIN — Linear sweep. The sweep variable is swept linearly from the starting to the ending value. *<increment value>* is the step size.

OCT — Sweep by octaves. The sweep variable is swept logarithmically by octaves. *<points value>* is the number of steps per octave.

DEC — Sweep by decades. The sweep variable is swept logarithmically by decades. *<points value>* is the number of steps per decade.

LIST — Use a list of values. In this case there are no start and end values. Instead, the numbers that follow the keyword LIST are the values that the sweep variable will be set to.

<sweep variable name> can be one of the following types:

Source: a name of an independent voltage or current source. During the sweep, the source's voltage or current is set to the sweep value.

Model parameter: a model type and model name followed by a model parameter name in parenthesis. The parameter in the model is set to the sweep value. The following model parameters cannot be (usefully) swept: **L** and **W** for the MOSFET device (use **LD** and **WD** as a work around), and any temperature parameters, such as **TC1** and **TC2** for the resistor, etc.

Temperature: use the keyword TEMP for *<sweep variable name>*. The temperature is set to the sweep value. For each value in the sweep, all the circuit components have their model parameters updated to that temperature.

Global Parameter: use the keyword PARAM, followed by the parameter name, for *<sweep variable name>*. During the sweep, the global parameter's value is set to the sweep value and all expressions are re-evaluated.

After the DC sweep is finished, *<sweep variable name>* is set back to the value it had before the sweep started.

.END — End of Circuit

General Form

.END

Example

```
.END
```

The .END statement marks the end of the circuit. All the data and commands must come before it. When the .END statement is reached, PSpice does all the specified analyses on the circuit.

There may be more than one circuit in an input file. Each circuit and its commands are marked by a .END statement. PSpice processes all the analyses for each circuit before going on to the next one. Everything is reset at the beginning of each circuit. Having several circuits in one file gives the same results as having them in separate files and running each one separately. However, all the simulation results go into one ".out" file and one ".dat" file. This is a convenient way to arrange a set of runs to be done overnight.

The last statement in an input file must be a .END statement.

.ENDS — End of Subcircuit Definition

General Form

.ENDS [*subcircuit name*]

Example

```
.ENDS
.ENDS OPAMP
```

The .ENDS statement marks the end of a subcircuit definition (started by a .SUBCKT statement). It is good practice to repeat the subcircuit name although this is not required.

.FOUR — Fourier Analysis

General Form

.FOUR *<frequency value>* [*no. harmonics value*] *<output variable>*

Examples

```
.FOUR 10kHz V(5) V(6,7) I(VSENS3)
.FOUR 60Hz 20 V(17)
```

Fourier analysis performs a decomposition into Fourier components of the result(s) of a transient analysis. This is accomplished by performing a Fourier integral on the selected outputs at evenly spaced time points. The time interval used is *<print step value>* in the .TRAN statement, or 1% of the *<final time value>* (TSTOP) if smaller, and a 2nd-order polynomial interpolation is used to calculate the output value used in the integration. A .FOUR statement requires a .TRAN statement.

<output variable> is an output variable of the same form as in a .PRINT statement or .PLOT statement for a transient analysis. The Fourier analysis is done by starting with the results of the transient analysis for the specified output variables. From these voltages/currents, the DC component, the fundamental, and the 2nd through 9th harmonics are calculated by default. Although, more harmonics can be specified. The fundamental frequency is *<frequency value>*. Not all of the transient results are used, only the interval from the end, back to 1/*<frequency value>* before the end is used. This means that the transient analysis must be at least 1/*<frequency value>* seconds long.

The .FOUR analysis does not require .PRINT, .PLOT, or .PROBE statements. The tabulated results are written to the output file (".out") as the transient analysis is done.

.INC Include File

General Form

.INC <"*file name*">

Examples

```
.INC "SETUP.CIR"
.INC "C:\LIB\VCO.CIR"
```

The .INC statement is used to insert the contents of another file. <"*file name*"> can be any character string which is a legal file name for your computer system.

Included files may contain any statements with the following exceptions: no title line is allowed (use a comment), .END statement (if present) marks only the end of included file, and .INC statement may be used (only up to 4 levels of "including").

Including a file is the same as simply bringing the file's text into the circuit file. Everything in the included file is actually read in: every model and subcircuit definition, even if not needed, takes up space in main memory (RAM).

.LIB Library File

General Form

.LIB ["*file name*"]

Examples

```
.LIB
.LIB "LINEAR.LIB"
.LIB "C:\LIB\BIPOLAR.LIB"
```

The .LIB statement is used to reference a model or subcircuit library in another file. <"*file name*"> can be any character string which is a legal file name for your computer system. The file extension **is not defaulted** to ".LIB". If you specify a file name, you **must include its extension.**

If "*file name*" is left off, all references will be done to the master library file, "nom.lib," which then references the individual library files. When a library file is referenced, PSpice will first search for the file in the current working directory, and then in the directory specified by the environment variable PSPICELIB. Follow the directions in the *Installation_Manual* to set the PSPICELIB environment variable on your system.

When any library is modified, PSpice creates an index file the first time the library is used. Any time thereafter, the look-up is noticeably faster. The index file is organized in a way which allows PSpice to locate a particular .MODEL or .SUBCKT quickly, in spite of how large the library file is. If you change a library file frequently, it is best not to reference that file from other libraries, since their index files will have to be reconstructed each time you change the library, and this may be time consuming.

Library files may contain comments, .MODEL statements, subcircuit definitions (including the .ENDS statement), .PARAM statements, and .LIB statements. No other statements are allowed. For further discussion of library files, see the "Device Libraries" section of the "Device Models, Subcircuits, and Libraries" chapter in the *Circuit Analysis User's Guide.*

.MODEL Model

General Form

.MODEL *<model name>* [AKO: *<reference model name>*]
+ *<model type>*
+ ([*<parameter name>* = *<value>* [*tolerance specification*]]*
+ [T_MEASURED=*<value>*] [[T_ABS=*<value>*] or
+ [T_REL_GLOBAL=*<value>*] or [T_REL_LOCAL=*<value>*]])

Examples

```
.MODEL RMAX RES (R=1.5 TC1=.02 TC2=.005)
.MODEL DNOM D (IS=1E-9)
.MODEL QDRIV NPN (IS=1E-7 BF=30)
.MODEL MLOAD NMOS(LEVEL=1 VTO=.7 CJ=.02pF)
.MODEL CMOD CAP (C=1 DEV 5%)
.MODEL DLOAD D (IS=1E-9 DEV .5% LOT 10%)
.MODEL RTRACK RES (R=1 DEV/GAUSS 1% LOT/UNIFORM 5%)
.MODEL QDR2 AKO:QDRIV NPN (BF=50 IKF=50m)
```

The .MODEL statement defines a set of device parameters which can be referenced by devices in the circuit. *<model name>* is the model name which devices use to reference a particular model. *<model name>* must start with a letter. It is good practice to make this the same letter as the device name (e.g., D for diode, Q for bipolar transistor, etc.) but this is not required.

The last example uses the AKO: (A Kind Of) syntax to reference the parameters of the model QDRIV in the third example. The model types of the current model and the AKO model must be the same. The value of each parameter of the referenced model is used unless overridden by the current model, e.g., the value of IS comes from QDRIV, but the values of BF and IKF come from the current definition. Parameter values or formulas are transferred, but not the tolerance specification. The referenced model may be in the main circuit file, accessed through a .INC statement, or may be in a library file.

<model type> is the device type and must be one of:

Figure B-2: Model Types

Model Type	Instance Name	Type of Device
CAP	Cxxx	Capacitor
IND	Lxxx	Inductor
RES	Rxxx	Resistor
D	Dxxx	Diode
NPN	Qxxx	NPN bipolar transistor
PNP	Qxxx	PNP bipolar transistor
LPNP	Qxxx	Lateral PNP bipolar transistor
NJF	Jxxx	N-channel junction FET
PJF	Jxxx	P-channel junction FET
NMOS	Mxxx	N-channel MOSFET
PMOS	Mxxx	P-channel MOSFET
GASFET	Bxxx	N-channel GaAs MESFET
CORE	Kxxx	Non-linear, magnetic core (transformer)
VSWITCH	Sxxx	Voltage-controlled switch
ISWITCH	Wxxx	Current-controlled switch
DINPUT	Nxxx	Digital input device (receive from digital)
DOUTPUT	Oxxx	Digital output device (transmit to digital)
UIO	Uxxx	Digital I/O model
UGATE	Uxxx	Standard gate
UTGATE	Uxxx	Tri-state gate
UEFF	Uxxx	Edge-triggered flip-flop
UGFF	Uxxx	Gated flip-flop
UWDTH	Uxxx	Pulse width checker
USUHD	Uxxx	Setup and hold checker
UDLY	Uxxx	Digital delay line
UADC	Uxxx	Multi-bit analog-to-digital converter
UDAC	Uxxx	Multi-bit digital-to-analog converter

Devices can reference models only of the correct type. A JFET can reference a model of types NJF or PJF, but not of type NPN. There can be more than one model of the same type in a circuit, although they must have different names.

.OP — Bias Point

General Form

.OP

Example

.OP

The .OP statement causes detailed information about the bias point to be printed. The bias point is calculated whether or not there is a .OP statement. Without a .OP statement, the only information about the bias point which is output is a list of the node voltages.

With a .OP statement, the small-signal (linearized) parameters of all the non-linear controlled sources and all the semiconductor devices are output.

The .OP statement controls output for the regular bias point only. The .TRAN statement controls output for the transient analysis bias point.

.PARAM — Parameter Definition

General Forms

.PARAM < *<name>* = *<value>* >*
.PARAM < *<name>* = { *<expression>* } >*

Examples

```
.PARAM VSUPPLY = 5V
.PARAM VCC = 12V, VEE = -12V
.PARAM BANDWIDTH = {100kHz/3}
.PARAM PI = 3.14159, TWO_PI = {2*3.14159}
.PARAM VNUM = {2*TWO_PI}
```

The keyword .PARAM is followed by a list of names with values. The values must be either constants or expressions. Constants (*<value>*) do not need "{" and "}". *<expression>* can contain constants or parameters. The parameters need not be previously defined. The .PARAM statement can be used inside a subcircuit definition to create local subcircuit parameters.

There are several predefined parameters:

TEMP	temperature (*reserved, not available this release*)
VT	thermal voltage (*reserved, not available this release*)
GMIN	shunt conductance for semiconductor *p-n* junctions

<name> cannot be one of these predefined parameters, nor can *<name>* be TIME or one of the .TEXT names.

Once defined, a parameter can be used in place of most numeric values in the circuit description. For example:

All model parameters.

All device parameters, such as AREA, L, NRD, and Z0. This includes IC= values on capacitors and inductors, but **not** the transmission line parameters NL and F, and **not the in-line** temperature coefficients for the resistor (of course, parameters can be used for the **TC1** and **TC2** resistor model parameters).

All independent voltage and current source (V and I device) parameters **except** for PWL values.

Not the E, F, G, and H device polynomial coefficient values and gain.

Values in .IC and .NODESET statements.

Parameters cannot be used in place of node numbers, nor can the values on analysis statements (.TRAN, .AC, etc.) be parameterized.

For more information see the *Circuit Analysis User's Guide.*

.PARAM statements can be in a library. PSpice will search libraries for parameters not defined in the circuit file, in the same way it searches for undefined models and subcircuits.

.PRINT Print

General Form

.PRINT[/DGTLCHG] *<analysis type>* [*output variable*]*

Examples

```
.PRINT DC  V(3)  V(2,3) V(R1) I(VIN) I(R2) IB(Q13) VBE(Q13)
.PRINT AC  VM(2) VP(2) VM(3,4) VG(5) VDB(5) IR(6) II(7)
.PRINT NOISE INOISE ONOISE DB(INOISE) DB(ONOISE)
.PRINT TRAN V(3) V(2,3) ID(M2) I(VCC)
.PRINT TRAN D(QA) D(QB) V(3) V(2,3)
.PRINT/DGTLCHG TRAN QA QB RESET
.PRINT TRAN V(3) V(R1) V([RESET])
```

The .PRINT statement allows results from DC, AC, noise, and transient analyses to be output in the form of tables, referred to as print tables. The .PRINT/DGTLCHG form is for digital output variables only. Values are printed for each output variable whenever one of the variables changes.

<analysis type> can be either DC, AC, NOISE, or TRAN, which can be output with .PRINT statements. Exactly one analysis type must be specified.

Following the analysis type is a list of the output variables. There is no limit to the number of output variables: the printout is split up depending on the width of the data columns (set with NUMDGT option) and the output width (set with WIDTH option).

The values of the output variables are printed as a table with each column corresponding to one output variable. The number of digits which are printed for analog values can be changed by the NUMDGT on the .OPTIONS statement.

The last example illustrates how to print a node which has a name rather than a number. The first item to print is a node voltage, the second item is the voltage across a resistor, and the third item to print is another node voltage, even though the second and third items both begin with the letter "R." The square brackets force the interpretation of names to mean node names.

An analysis may have any number of .PRINT statements.

.PROBE Probe

General Form

.PROBE[/CSDF][*output variable*]*

Examples

```
.PROBE
.PROBE V(3) V(2,3) V(R1) I(VIN) I(R2) IB(Q13) VBE(Q13)
.PROBE/CSDF
.PROBE V(3) V(R1) V([RESET])
.PROBE D(QBAR)
```

The .PROBE statement writes the results from DC, AC, and transient analyses to a data file named "probe.dat" for use by the Probe waveform analyzer. See the "Waveform Analysis" chapter in the *Circuit Analysis User's Guide* for a description of Probe and using the "probe.dat" file.

The first form (with no output variables) writes all the node voltages and all the device currents to the data file. The list of device currents written is the same as the device currents allowed as output variables, as described below.

The second form writes only those output variables specified to the data file. Note that unlike the .PRINT and .PLOT statements, there is no analysis name before the output variables. Also, the number of output variables is not restricted to 8. See the table below for a description of the possible output variables. This form is intended for users without a fixed disk who need to limit the size of the "probe.dat" file written.

The third example creates a data file in a text format using the Common Simulation Data File (CSDF) format, not a binary format. This format is primarily used for transfers between different computer families. See the "Waveform Analysis" chapter in the *Circuit Analysis User's Guide* for more information on the use of text files with Probe.

The next example illustrates how to specify a node which has a name rather than a number. The first item to output is a node voltage, the second item is the voltage across a resistor, and the third item to output is another node voltage, even though the second and third items both begin with the letter "R." The square brackets force the interpretation of names to mean node names.

The last example only writes the output at digital node QBAR to the data file.

<u>Output Variables</u>

This section describes the types of output variables allowed in the .PRINT, .PLOT, and .PROBE statements. Each .PRINT or .PLOT may have up to 8 output variables. This format is similar to that used when calling up waveforms while running Probe.

DC Sweep and Transient Analysis

For DC sweep and transient analysis, these are the available output variables:

Figure B-3: Output Variables for DC Sweep and Transient Analysis

General Form	Meaning
V(<*node*>)	voltage at node
V(<+ *node*> , <- *node*>)	V across + and - nodes
V(<*name*>)	V across two-term. device
V*x*(<*name*>)	V at a non-grounded terminal
V*z*(<*name*>)	V at one end of a transmission line
I(<*name*>)	I through <*name*>
I*x*(<*name*>)	I into terminal
I*z*(<*name*>)	I at one end of a transmission line
D(<*name*>)	digital value of <*name*> (a digital node). These values are available for transient and DC analysis only. For the .PRINT/DGTLCHG statement the "D( )" is optional.

Figure B-4: Output Variable Examples for DC Sweep and Transient Analysis

Examples	Meaning
V(3)	V between node 3 and ground
V(3,2)	voltage between nodes 3 and 2
V(R1)	voltage across resistor R1
VB(Q3)	voltage between base of transistor Q3 and ground
VGS(M13)	gate-source voltage of M13
VA(T2)	V at port A of T2
I(D5)	I through diode D5
IG(J10)	current into gate of J10
D(QA)	the value of digital node QA

For the V(<*name*>) and I(<*name*>) forms, where <*name*> must be the name of a two-terminal device, the devices are:

Figure B-5: Two-Terminal Devices

Character ID	Device
C	capacitor
D	diode
E	voltage-controlled voltage source
F	current-controlled current source
G	voltage-controlled current source
H	current-controlled voltage source)
I	independent current source
L	inductor
R	resistor
S	voltage-controlled switch
V	independent voltage source
W	current-controlled switch

For the V*x*(<*name*>), V*xy*(<*name*>), and I*x*(<*name*>) forms, where <*name*> must be the name of a three or four-terminal device and *x* and *y* must each be a terminal abbreviation, the devices and the terminals are:

Figure B-6: Three and Four-Terminal Devices and Terminals

Device Type	Terminal Abbreviation
B (GaAs MESFET)	D (drain) G (gate) S (source)
J (Junction FET)	D (drain) G (gate) S (source)
M (MOSFET)	D (drain) G (gate) S (source) B (bulk, substrate)
Q (Bipolar transistor)	C (collector) B (base) E (emitter) S (substrate)

For the V*z*(<*name*>) and I*z*(<*name*>) forms, <*name*> must be the name of a transmission line (T device) and *z* must be "A" or "B." "A" means port A (the first two nodes) and "B" means port B (the last two nodes).

AC Analysis

For AC analysis, the output variables listed in the preceding section are augmented by adding a suffix. These are the available suffixes:

Figure B-7: Output Variables for AC Analysis

Suffix	Meaning
none	magnitude
M	magnitude
DB	magnitude in decibels
P	phase
G	group delay (-*d*PHASE/*d*FREQUENCY)
R	real part
I	imaginary part

Figure B-8: Output Variable Examples for AC Analysis

Examples	Meaning
V(2,3)	Magnitude of complex voltage across nodes 2 & 3
VM(2)	Magnitude of V at node 2
VDB(R1)	db magnitude of V across R1

.STEP Parametric Analysis

General Forms

.STEP [*linear sweep type*] *<sweep variable name>*
+ *<start value> <end value> <increment value>*

.STEP *<logarithmic sweep type> <sweep variable name>*
+ *<start value> <end value> <points value>*

.STEP *<sweep variable name>* LIST *<value>**

Examples

```
.STEP VCE 0V 10V .5V
.STEP LIN I2 5mA -2mA 0.1mA
.STEP RES RMOD(R) 0.9 1.1 .001
.STEP DEC NPN QFAST(IS) 1E-18 1E-14 5
.STEP TEMP LIST 0 20 27 50 80 100
.STEP PARAM CenterFreq 9.5kHz 10.5kHz 50Hz
```

The following examples illustrate two ways of stepping a resistor from 30 to 50 ohms in steps of 5 ohms.

This example uses a global parameter:

```
.PARAM RVAL = 1
R1 1 2 {RVAL}
.STEP PARAM RVAL 30,50,5
```

RVAL is the global parameter and PARAM is the keyword used by the .STEP statement when using a global parameter.

The following example steps the resistor model parameter R:

```
R1 1 2 RMOD 1
.MODEL RMOD RES(R=30)
.STEP RES RMOD(R) 30,50,5
```

(**Note:** Do not uset R={30}.)

RMOD is the model name, RES is the sweep variable name (a model type), and R is the parameter within the model to step. To step the value of the resistor, the line value of the resistor is multiplied by the R parameter value to achieve the final resistance value, that is

final resistor value = line resistor value · R

Therefore, if you set the line value of the resistor to 1 ohm, the final resistor value is 1 · R or R. Thus, stepping R from 30 to 50 ohms will then step the resistor value from 1 · 30 ohms to 1 · 50 ohms.

In both examples, all of the ordinary analyses (.DC, .AC, .TRAN, etc.) are done for each step.

The .STEP statement causes a parametric sweep to be performed on *<sweep variable name>*, for all of the analyses of the circuit. .STEP is at the same "level" as the .TEMP command: all of the ordinary analyses (.DC, .AC, .TRAN, etc.) are done for each step. Once all the runs have finished, an entire .PRINT table or .PLOT plot for each value of the sweep will be output (Probe allows nested sweeps to be displayed as a family of curves), just as for the .TEMP or .MC commands. The first form, and the first three examples, are for doing a linear sweep. The second form, and the fourth example, are for doing a logarithmic sweep. The third form, and the fifth example, are for using a list of values for the sweep variable.

<start value> may be greater or less than *<end value>*: that is, the sweep may go in either direction. *<increment value>* and *<points value>* **must be greater than zero.**

The sweep can be linear, logarithmic, or a list of values. For [*linear sweep type*], the keyword LIN is optional, but either OCT or DEC must be specified for the *<logarithmic sweep type>*. The sweep types are:

LIN	Linear sweep. The sweep variable is swept linearly from the starting to the ending value. *<increment value>* is the step size.
OCT	Sweep by octaves. The sweep variable is swept logarithmically by octaves. *<points value>* is the number of steps per octave.
DEC	Sweep by decades. The sweep variable is swept logarithmically by decades. *<points value>* is the number of steps per decade.

LIST — Use a list of values. In this case there are no start and end values. Instead, the numbers that follow the keyword LIST are the values that the sweep variable will be set to. **Note:** The values must be in either ascending or descending order.

<sweep variable name> can be one of the following types:

Source: a name of an independent voltage or current source. During the sweep, the source's voltage or current is set to the sweep value.

Model parameter: a model type and model name followed by a model parameter name in parenthesis. The parameter in the model is set to the sweep value.

Temperature: use the keyword TEMP for *<sweep variable name>*. The temperature is set to the sweep value. For each value in the sweep, all the circuit components have their model parameters updated to that temperature.

Global Parameter: use the keyword PARAM, followed by the parameter name, for *<sweep variable name>*). During the sweep, the global parameter's value is set to the sweep value and all expressions are re-evaluated.

The .STEP statement is similar to the .DC statement and immediately raises the question of what happens if both .STEP and .DC try to set the same value. The same question can come up with Monte Carlo analysis. The answer is that **this is disallowed**: no two analyses (.STEP, .TEMP, .MC, .WCASE, and .DC) can try to set the same value. This is flagged as an error during read-in and no analyses are done.

The .STEP command provides the capability to look at the response of a circuit as a parameter varies. For example, how does the center frequency of a filter shift as a capacitor varies? With .STEP you can vary that capacitor and then see a family of AC waveforms that show the variation. Similar comments apply to looking at, for example, propagation delay in transient analysis.

.SUBCKT — Subcircuit Definition

General Form

.SUBCKT*<name>* [*node*]*
+ [OPTIONAL: < *<interface node>* = *<default value>* >*]
+ [PARAMS: < *<name>* = *<value>* >*]
+ [TEXT: < *<name>* = *<text value>* >*]

Examples

```
.SUBCKT OPAMP 1 2 101 102 17
.SUBCKT FILTER INPUT, OUTPUT PARAMS: CENTER=100kHz, WIDTH=10kHz
.SUBCKT PLD IN1 IN2 IN3 OUT1
+  PARAMS: MNTYMXDLY=0 IO_LEVEL=0
+  TEXT: JEDEC FILE="PROG.JED"
.SUBCKT  74LS00  A B Y
+  OPTIONAL: DPWR=$G_DPWR DGND=$G_DGND
+  PARAMS: MNTYMXDLY=0 IO_LEVEL=0
```

The .SUBCKT statement begins the definition of a subcircuit. The definition is ended with a .ENDS statement. All the statements between .SUBCKT and .ENDS are included in the definition. Whenever the subcircuit is called, by an X statement, all the statements in the definition replace the calling statement.

<name> is the subcircuit's name and is used by an X statement to reference the subcircuit.

[*node*]* is an optional list of nodes (pins). There must be the same number of nodes in the subcircuit calling statements as in its definition. When the subcircuit is called, the actual nodes (the ones in the calling statement) replace the argument nodes (the ones in the defining statement). **Do not** use 0 ("zero") in this node list: that is reserved for global "ground" node.

The **Optional:** keyword allows you to specify one or more optional nodes (pins) in the subcircuit definition. The optional nodes are stated as a pair consisting of an interface node and its default value. If an optional node is not specified in a subcircuit call (X statement), its default value is used inside the subcircuit; otherwise, the value specified in the subcircuit call is used.

This feature is particularly useful when specifying power supply nodes, because the same nodes are normally used in every device. This makes the subcircuits easier to use because the same two nodes do not have to be specified in each subcircuit call. This method is used in the libraries provided with the Digital Simulation feature.

.TEMP — Temperature

General Form

.TEMP *<temperature value>**

Examples

```
.TEMP 125
.TEMP 0 27 125
```

The .TEMP statement sets the temperature at which all analyses are done. The temperatures are in degrees Centigrade. If more than one temperature is given, then all analyses are done for each temperature.

It is assumed that the model parameters were measured or derived at the nominal temperature, TNOM (27°C by default). See the .OPTIONS statement (page 60) for setting TNOM.

.TRAN — Transient Analysis

General Form

.TRAN[/OP] *<print step value>* *<final time value>*
+[*no-print value* [*step ceiling value*]] [UIC]

Examples

```
.TRAN 1ns 100ns
.TRAN/OP 1ns 100ns 20ns UIC
.TRAN 1ns 100ns 0ns .1ns
```

The .TRAN statement causes a transient analysis to be performed on the circuit. The transient analysis calculates the circuit's behavior over time, starting at TIME=0 and going to *<final time value>*.

The variables TSTEP and TSTOP, which are used in defaulting some waveform parameters, are set by the .TRAN command. TSTEP is *<print step value>* and TSTOP is *<final time value>*. The .TRAN command can be anywhere in the circuit file; it need not come after the voltage source.

The transient analysis uses an internal time step which is adjusted as the analysis proceeds. Over intervals where there is little activity, the internal time step is increased and during busy intervals it is decreased. *<print step value>* is the time interval used for printing, plotting, (.PRINT or .PLOT), or performing a Fourier integral on the results of the transient analysis. Since the results are computed at different times than they are printed, a 2nd-order polynomial interpolation is used to obtain the printed values. This applies only to .PRINT, .PLOT, and .FOUR outputs and does not affect Probe.

The transient analysis always starts at TIME=0. However, it is possible to suppress output of a portion of the analysis. [*no-print value*] is the amount of time from TIME=0 which is not printed, plotted, or given to Probe.

Sometimes one is concerned about the size of the internal time step. The default ceiling on the internal time step is *<final time value>*/50 (**it is** *<print step value>* only if there are no charge storage elements, inductances, or capacitances in the circuit). [*step ceiling value*] allows a ceiling smaller or larger than the print interval to be put on the internal time step.

Prior to doing the transient analysis, PSpice computes a bias point for the circuit separate from the regular bias point. This is done because the independent sources can have different values at the start of a transient analysis than their DC value. Normally, only the node voltages are printed for the transient analysis bias point. However, the "/OP" suffix (on .TRAN) will cause the same detailed printing of the bias point that the .OP statement causes for the regular bias point.

If the keyword UIC (Use Initial Conditions) is put at the end of the .TRAN statement, the calculation of the bias point is skipped. This option is used with the IC= specification for capacitors and inductors. See the "Standard Analyses; Setting Initial Conditions" section of the "Analysis Specification" chapter in the *Circuit Analysis User's Guide* for more information on setting initial conditions.

.PRINT, .PLOT, .FOUR, or .PROBE statements must be used to get the results of the transient analysis.

.WATCH Watch Analysis Results

General Form

.WATCH [DC][AC][TRAN]
+ [*<output variable>* [*<lower limit value>*,*<upper limit value>*]]*

Examples

```
.WATCH DC V(3) (-1V,4V) V(2,3) V(R1)
.WATCH AC VM(2) VP(2) VMC(Q1)
```

```
.WATCH TRAN VBE(Q13) (0V,5V) ID(M2) I(VCC) (0,500mA)
.WATCH DC V([RESET]) (2.5V,10V)
```

The .WATCH statement allows results from DC, AC, and transient analyses to be output to the screen while the simulation is running. Up to three output variables can be seen on the display at one time. More than three variables can be specified, but they will not all be displayed.

DC, AC, and TRAN are the analyses types which can be output with the .WATCH statement. Exactly one analysis type must be specified per .WATCH statement, and there can be a .WATCH statement for each analysis type in the circuit.

Following the analysis type is a list of the output variables with optional value ranges. A maximum of eight output variables are allowed on a single .WATCH statement.

The optional value range specifies the normal operating range of that particular output variable. If the range is exceeded during the simulation, the simulator will beep and pause. At this point, the simulation can be aborted or continued. If continued, the check for that output variable's boundary condition will be eliminated. Each output variable can have its own value range.

The first example above displays three output variables on the screen. The first variable, V(3), has an operating range set from -1 V to 4 V. If during the simulation the voltage at node 3 exceeds 4 volts, the simulation will pause. If the simulation is allowed to proceed, and node 3 continues to rise in value, the simulation will not be interrupted. However, if the simulation is allowed to continue and V(3) falls below -1 V, the simulation would pause again because a new boundary condition was exceeded.

* Comment

General Form

* [*any text*]

Example

```
* This is an example of a comment
```

A statement beginning with "*" is a comment line and has no effect. The use of comment statements throughout the input is recommended. It is good practice to place a comment just before a subcircuit definition to identify the nodes, for example

```
*                    +IN -IN V+ V- +OUT -OUT
.SUBCKT OPAMP 100   101  1  2   200  201
```

or to identify major blocks of circuitry.

; In-line Comment

General Form

circuit file text ;[*any text*]

Examples

```
R13 6 8 10K ; feedback resistor
C3 15 0 .1U ; decouple supply
```

A ";" is treated as the end of a line: PSpice moves on to the next line in the circuit file. The text after the ";" is a comment and has no effect. The use of comments throughout the input is recommended. This type of comment can also replace comment lines, which must start with "*" in the first column.

C

Abridged List of Analog Devices and Libraries

Figure C-1: Analog Device Summary

Device Class	Type	Description	Libraries
Passive	C	Capacitor	none
	K	Inductor coupling	magnetic.lib
	L	Inductor	none
	R	Resistor	none
	T	Transmission line	none
	Quartz Crystal†		xtal.lib
Independent Sources	I	Current source	none
	V	Voltage source	none
Controlled Sources	E	Voltage-controlled voltage source	none
	F	Current-controlled current source	none
	G	Voltage-controlled current source	none
	H	Current-controlled voltage source	none
Ideal Switches	S	Voltage-controlled switch	none
	W	Current-controlled switch	none
Semiconductor	B	GaAsFET	none
	D	Diode	diode.lib
			europe.lib
	J	JFET	jfet.lib
	M	MOSFET	motormos.lib
			polyfet.lib
	Q	Bipolar transistor	pwrmos.lib
			bipolar.lib
			europe.lib
	Opamp and		powerbjt.lib
	Comparator†		adv_lin.lib

Figure C-1: Analog Device Summary (Continued)

Device Class	Type	Description	Libraries
Semiconductor	Opamp and Comparator (cont.)†		anlg_dev.lib apex.lib burr_brn.lib comlinr.lib elantec.lib harris.lib linear.lib lin_tech.lib nat_semi.lib tex_inst.lib
	SCR†		thyristr.lib
	triac†		thyristr.lib
	UJT†		thyristr.lib
	opto-coupler†		opto.lib
	555 timer†		misc.lib
	pulse-width modulator†		swit_reg.lib
	averaging power supply†		swit_rav.lib
Miscellaneous	X	Subcircuits (generic form)	(any)
	filters†	Biquad stages, LC ladder branches	filtsub.lib

† Subcircuit device types referenced via the X device type declaration.

C Capacitor

General Form

C<*name*> <(+) *node*> <(-) *node*> [*model name*] <*value*>
+ [IC=<*initial value*>]

Examples

```
CLOAD 15 0 20pF
C2 1 2 .2E-12 IC=1.5V
CFDBCK  3 33 CMOD 10pF
```

Model Form

.MODEL <*model name*> CAP [*model parameters*]

Figure C-2: Capacitor Model Parameters

Model Parameters(see .MODEL statement)	Description	Units	Default
C	Capacitance multiplier		1
VC1	Linear voltage coefficient	$volt^{-1}$	0
VC2	Quadratic voltage coefficient	$volt^{-2}$	0
TC1	Linear temperature coefficient	$°C^{-1}$	0
TC2	Quadratic temperature coefficient	$°C^{-2}$	0
T_MEASURED	Measured temperature	°C	
T_ABS	Absolute temperature	°C	
T_REL_GLOBAL	Relative to current temperature	°C	
T_REL_LOCAL	Relative to AKO model temperature	°C	

The (+) and (-) nodes define the polarity meant when the capacitor has a positive voltage across it. Positive current flows from the (+) node through the capacitor to the (-) node.

If [*model name*] is left out then <*value*> is the capacitance in farads.

If [*model name*] is specified, then the capacitance is given by the formula

<*value*>·C·(1+**VC1**·V+**VC2**·V^2)·(1+**TC1**·(T-Tnom)+**TC2**·$(T-Tnom)^2$)

where <*value*> is normally positive (though it can be negative, but **not** zero). "Tnom" is the nominal temperature (set with TNOM option).

D Diode

General Form

D<*name*> <(+) *node*> <(-) *node*> <*model name*> [*area value*]

Examples

```
DCLAMP 14  0 DMOD
D13 15 17 SWITCH 1.5
```

Model Form

.MODEL <*model name*> D [*model parameters*]

Figure C-3: Diode Model Parameters

Model Parameters (see .MODEL statement)	Description	Unit	Default
IS	Saturation current	amp	1E-14
N	Emission coefficient		1
ISR	Recombination current parameter	amp	0
NR	Emission coefficient for ISR		2
IKF	High-injection "knee" current	amp	infinite
BV	Reverse breakdown "knee" voltage	volt	infinite
IBV	Reverse breakdown "knee" current	amp	1E-10
NBV	Reverse breakdown ideality factor		1
IBVL	Low-level reverse breakdown "knee" current	amp	0
NBVL	Low-level reverse breakdown ideality factor		1
RS	Parasitic resistance	ohm	0
TT	Transit time	sec	0
CJO	Zero-bias *p-n* capacitance	farad	0
VJ	*p-n* potential	volt	1
M	*p-n* grading coefficient		0.5
FC	Forward-bias depletion capacitance coefficient		0.5
EG	Bandgap voltage (barrier height)	eV	1.11
XTI	IS temperature exponent		3
TIKF	IKF temperature coefficient (linear)	$°C^{-1}$	0
TBV1	BV temperature coefficient (linear)	$°C^{-1}$	0
TBV2	BV temperature coefficient (quadratic)	$°C^{-2}$	0
TRS1	RS temperature coefficient (linear)	$°C^{-1}$	0
TRS2	RS temperature coefficient (quadratic)	$°C^{-2}$	0
KF	Flicker noise coefficient		0
AF	Flicker noise exponent		1
T_MEASURED	Measured temperature	°C	
T_ABS	Absolute temperature	°C	
T_REL_GLOBAL	Relative to current temperature	°C	
T_REL_LOCAL	Relative to AKO model temperature	°C	

E Voltage-Controlled Voltage Source

General Forms

E<*name*> <(+) *node*> <(-) *node*> (+) *controlling node*> <(-) *controlling node*> <*gain*>

E<*name*> <(+) *node*> <(-) *node*> POLY(<*value*>)
+ < <(+) *controlling node*> <(-) *controlling node*> >*
+ < <*polynomial coefficient value*> >*

E<*name*> <(+) <*node*> <(-) *node*> VALUE = { <*expression*> }

E<*name*> <(+) <*node*> <(-) *node*> TABLE { <*expression*> =
+ < <*input value*>,<*output value*> >*

E<*name*> (+) *node*> <(-) *node*> LAPLACE { <*expression*> } =
+ { <*transform*> }

E<*name*> <(+) *node*> <(-) *node*> FREQ { <*expression*> } =
+ < <*frequency value*>,<*magnitude value*>,<*phase value*> >*

E<*name*> <(+) *node*> <(-) *node*> CHEBYSHEV { <*expression*> } =
+ <[LP] [HP] [BP] [BR]>,<*cutoff frequencies*>*,<*attenuation*>*

Examples

```
EBUFF 1 2 10 11 1.0
EAMP 13 0 POLY(1) 26  0 0 500
ENONLIN 100 101 POLY(2) 3 0  4 0  0.0 13.6 0.2 0.005
ESQROOT   5   0 VALUE = {5V*SQRT(V(3,2))}
ET2 2 0 TABLE {V(ANODE,CATHODE)} = (0,0) (30,1)
ERC 5 0 LAPLACE {V(10)} = {1/(1+.001*s)}
ELOWPASS 5 0 CHEBYSHEV {V(10)} = LP 800 1.2K .1dB 50dB
```

The first form and the first two examples apply to the linear case. The second form and the last example are for the non-linear case. POLY(<*value*>) specifies the number of dimensions of the polynomial. The number of pairs of controlling nodes must be equal to the number of dimensions.

The (+) and (-) nodes are the output nodes. Positive current flows from the (+) node through the source to the (-) node. The <(+) *controlling node*> and <(-) *controlling node*> are in pairs and define a set of controlling voltages. A particular node may appear more than once, and the output and controlling nodes need not be different. The TABLE form has a maximum size of 2048 input/output value pairs.

Chebyshev filters have two attenuation values, given in dB, which specify the pass band ripple and the stop band attenuation. They may be given in either order, but must appear after all of the cutoff frequencies have been given. Low pass (LP) and high pass (HP) have two cutoff frequencies, specifying the pass band and stop band edges, while band pass (BP) and band reject (BR) filters have four. Again, these may be given in any order.

F Current-Controlled Current Source

General Forms

F<*name*> <(+) *node*> <(-) *node*>
+ <*controlling V device name*> <*gain*>

F<*name*> <(+) *node*> <(-) *node*> POLY(<*value*>)
+ <*controlling V device name*>*
+ < <*polynomial coefficient value*> >*

Examples

```
FSENSE 1 2 VSENSE 10.0
FAMP 13 0 POLY(1) VIN 0 500
FNONLIN 100 101 POLY(2) VCNTRL1 VCINTRL2 0.0 13.6 0.2 0.005
```

The first form and the first two examples apply to the linear case. The second form and the last example are for the non-linear case. POLY(<*value*>) specifies the number of dimensions of the polynomial. The number of controlling voltage sources must be equal to the number of dimensions.

The (+) and (-) nodes are the output nodes. A positive current will flow from the (+) node through the source to the (-) node. The current through the controlling voltage source determines the output current. The controlling source must be an independent voltage source (V device), although it need not have a zero DC value.

For the linear case, there must be one controlling voltage source and its name is followed by the gain. For the non-linear case (POLY), see the "Analog Behavioral Modeling" chapter in the *Circuit Analysis User's Guide* for describing the controlling polynomial.

Note: Expressions **cannot** be used for linear and polynomial coefficient values in a current-controlled current source device statement.

G Voltage-Controlled Current Source

General Forms

G<*name*><(+) *node*> <(-) *node*>
+ <(+) *controlling node*> <(-) *controlling node*>
+ <*transconductance*>

G<*name*> <(+) *node*> <(-) *node*> POLY(<*value*>)
+ < <(+) *controlling node*> <(-) *controlling node*> >*
+ <*polynomial coefficient value*>*

G<*name*> <(+) *node*> <(-) *node*> VALUE = { <*expression*> }

G<*name*> <(+) *node*> <(-) *node*> TABLE { <*expression*> } =
+ < <*input value*>,<*output value*> >*

G<*name*> <(+) *node*> <(-) *node*> LAPLACE { <*expression*> } =
+ { <*transform*> }

G<*name*> <(+) *node*> <(-) *node*> FREQ { <*expression*> } =
+ < <*frequency value*>,<*magnitude value*>,<*phase value*> >*

G<*name*> <(+) *node*> <(-) *node*> CHEBYSHEV { <*expression*> } =
+ <*type*>,<*cutoff frequencies*>*,<*attenuation*>*

Examples

```
GBUFF 1 2 10 11  1.0
GAMP 13 0 POLY(1) 26  0 0 500
GNONLIN 100 101 POLY(2) 3 0 4 0 0.0 13.6 0.2 0.005
GPSK 11 6 VALUE = {5MA*SIN(6.28*10kHz*TIME+V(3))}
GT ANODE CATHODE VALUE = {200E-6*PWR(V(1)*V(2),1.5)}
GLOSSY 5 0 LAPLACE {V(10)} = {exp(-sqrt(C*s*(R+L*s)))}
```

Note: The linear and POLY forms are part of the basic PSpice, and the VALUE, TABLE, LAPLACE, FREQ, and CHEBYSHEV forms are part of the Analog Behavioral Modeling feature. Please see the "Analog Behavioral Modeling" chapter in the *Circuit Analysis User's Guide* for more information on using these forms.

The first form and the first two examples apply to the linear case. The second form and the last example are for the non-linear case. POLY(<*value*>) specifies the number of dimensions of the polynomial. The number of pairs of controlling nodes must be equal to the number of dimensions.

The (+) and (-) nodes are the output nodes. A positive current flows from the (+) node through the source to the (-) node. The <(+) *controlling node*> and <(-) *controlling node*> are in pairs and define a set of voltages. A particular node may appear more than once, and the output and controlling nodes need not be different. The TABLE form has a maximum size of 2048 input/output value pairs.

For the linear case, there are two controlling nodes and these are followed by the transconductance. For the non-linear case (POLY), see the "Analog Behavioral Modeling" chapter in the *Circuit Analysis User's Guide* for describing the controlling polynomial.

Note: Expressions **cannot** be used for linear and polynomial coefficient values in a voltage-controlled current source device statement.

H Current-Controlled Voltage Source

General Forms

H<*name*> <(+) *node*> <(-) *node*>
+ <*controlling V device name*> <*transresistance*>

H<*name*> <(+) *node*> <(-) *node*> POLY(<*value*>)
+ <*controlling V device name*>*
+ <*polynomial coefficient value*>*

Examples

```
HSENSE 1 2 VSENSE 10.0
HAMP 13 0 POLY(1) VIN 0 500
HNONLIN 100 101 POLY(2) VCNTRL1 VCINTRL2 0.0 13.6 0.2 0.005
```

The first form and the first two examples apply to the linear case. The second form and the last example are for the non-linear case. POLY(<*value*>) specifies the number of dimensions of the polynomial. The number of controlling voltage sources must be equal to the number of dimensions.

The (+) and (-) nodes are the output nodes. Positive current flows from the (+) node through the source to the (-) node. The current through the controlling voltage source determines the output voltage. The controlling source must be an independent voltage source (V device), though it need not have a zero DC value.

For the linear case, there must be one controlling voltage source and its name is followed by the transresistance. For the non-linear case (POLY), see the "Analog Behavioral Modeling" chapter in the *Circuit Analysis User's Guide* for describing the controlling polynomial.

Note: Expressions **cannot** be used for linear and polynomial coefficient values in a current-controlled voltage source device statement.

I Independent Current Source and Stimulus

General Form

I<*name*> <(+) *node*> <(-) *node*>
\+ [[DC] <*value*>]
\+ [AC <*magnitude value*> [*phase value*]]
\+ [*transient specification*]

Examples

```
IBIAS 13 0 2.3mA
IAC 2 3 AC .001
IACPHS 2 3 AC .001 90
IPULSE 1 0 PULSE(-1mA 1mA 2ns 2ns 2ns 50ns 100ns)
I3 26 77 DC .002 AC 1 SIN(.002 .002 1.5MEG)
```

This element is a current source. Positive current flows from the (+) node through the source to the (-) node: in the first example, IBIAS drives node 13 to have a **negative** voltage. The default value is zero for the DC, AC, and transient values. None, any, or all of the DC, AC, and transient values may be specified. The AC phase value is in degrees.

If present, the [*transient specification*] must be one of:

EXP (<*parameters*>)	for an exponential waveform
PULSE (<*parameters*>)	for a pulse waveform
PWL (<*parameters*>)	for a piecewise linear waveform
SFFM (<*parameters*>)	for a frequency-modulated waveform
SIN (<*parameters*>)	for a sinusoidal waveform

The variables TSTEP and TSTOP, which are used in defaulting some waveform parameters, are set by the .TRAN command. TSTEP is *<print step value>* and TSTOP is *<final time value>*. The .TRAN command can be anywhere in the circuit file; it need not come after the voltage source.

General Form

EXP (<i1> <i2> <td1> <tc1> <td2> <tc2>)

Example

IRAMP 10 5 EXP(1 5 1 .2 2 .5)

Figure C-4: Independent Current Source and Stimulus Exponential Waveform Parameters

Parameters	Description	Units	Default
<i1>	Initial current	amp	none
<i2>	Peak current	amp	none
<td1>	Rise delay	sec	0
<tc1>	Rise time constant	sec	TSTEP
<td2>	Fall delay	sec	<td1>+TSTEP
<tc2>	Fall time constant	sec	TSTEP

Figure C-5: EXP Current Waveform

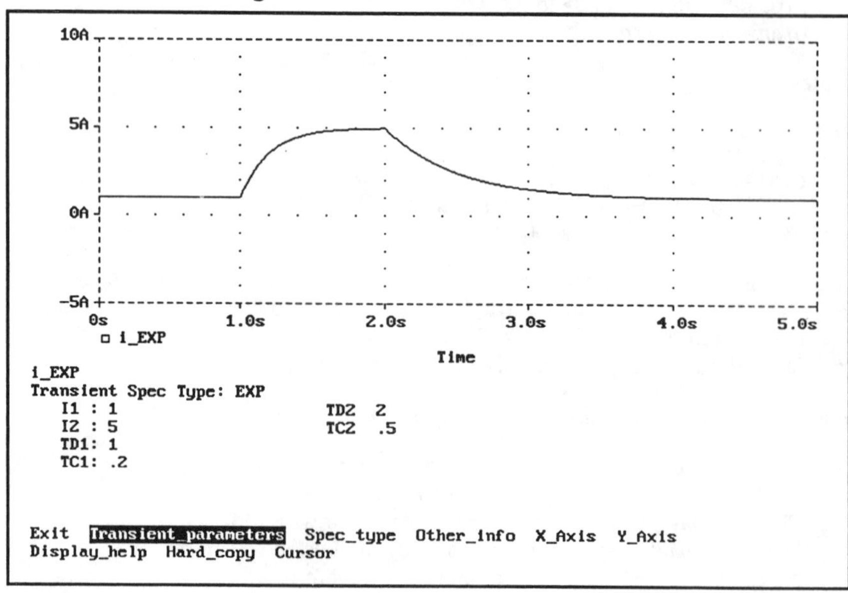

General Form

PULSE (<i1> <i2> <td> <tr> <tf> <pw> <per>)

Example

ISW 10 5 PULSE(1A 5A 1sec .1sec .4sec .5sec 2sec)

Figure C-6: Independent Current Source and Stimulus Pulse Waveform Parameters

Parameters	Description	Units	Default
<i1>	Initial current	amp	none
<i2>	Pulsed current	amp	none
<td>	Delay	sec	0
<tr>	Rise time	sec	TSTEP
<tf>	Fall time	sec	TSTEP
<pw>	Pulse width	sec	TSTOP
<per>	Period	sec	TSTOP

Figure C-7: PULSE Current Waveform

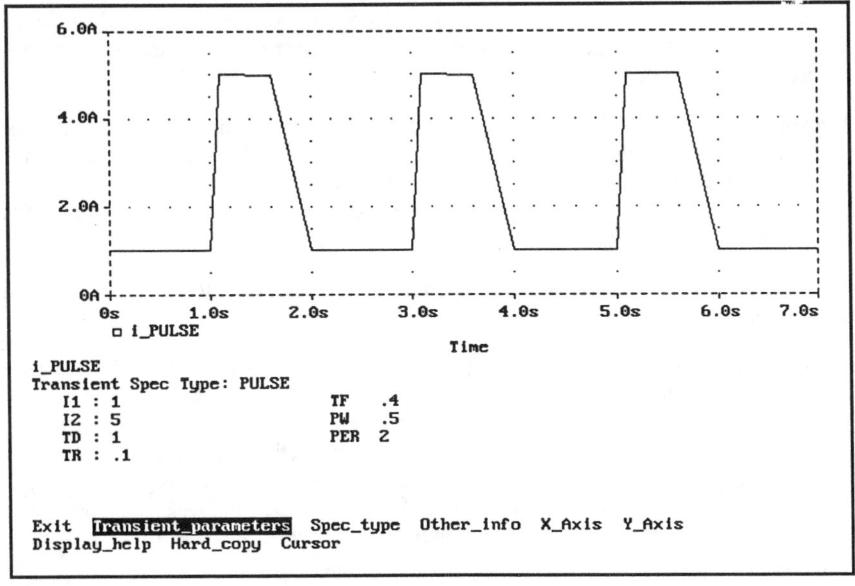

General Form

PWL [*<time_scale_factor>*] [*<value_scale_factor>*](*<tn>* *<in>*)*
+ [[REPEAT FOR *<n>*] (*<tn>* *<in>*)* [ENDREPEAT]]*
+ [[REPEAT FOREVER] (*<tn>* *<in>*)* [ENDREPEAT]]*

or with time, value pairs (*<tn>*, *<in>*) contained within a file:

PWL [*<time_scale_factor>*] [*<value_scale_factor>*] FILE*<file name>*

Examples

```
I3 10 5 PWL(O O 1 OA 1.2 5A 1.4 2A 2 4A 3 1A)
```

Figure C-8: Independent Current Source and Stimulus Piecewise Linear Waveform Parameters

Parameters	Description	Units	Default
<tn>	Time at corner	sec	none
<in>	Current at corner	amp	none
<n>	number of repetitions	positive integer	none

The PWL form describes a piecewise linear waveform. Each pair of time-current values specifies a corner of the waveform. Up to 3995 pairs may be used for the IBM-PC and the Macintosh. The current at times between corners is the linear interpolation of the currents at the corners. This behavior is shown in Figure C-9.

The keywords *<time_scale_factor>* and/or *<value_scale_factor>* may be coded immediately after the PWL keyword to indicate that the time and/or current value pairs are to be multiplied by the appropriate scale factor. These scale factors may be expressions. If they are expressions, they are evaluated once per outer simulation loop, and thus should be composed of expressions not containing references to voltages or currents.

The values in the time-current pairs (*<tn>* *<in>*) may be expressions with the same restrictions as the scaling keywords.

All the time-current pairs may be replaced by FILE*<file name>*. The file named "*<file name>*" will be read to supply the (*<tn>* *<in>*) pairs. The specified file will be a text file containing the time-current pairs. The contents of this file are read by the same parser that reads the circuit file. Thus, engineering units (e.g., 10us) will be correctly interpreted. Note that the continuation + signs in the first column are unnecessary and are discouraged. A typical file may be created by editing an existing PWL specification, replacing all + signs with blanks (to avoid unintentional +time). Only numbers (with possible units attached) may appear in the file; expressions for values are not allowed.

Figure C-9: PWL Current Waveform

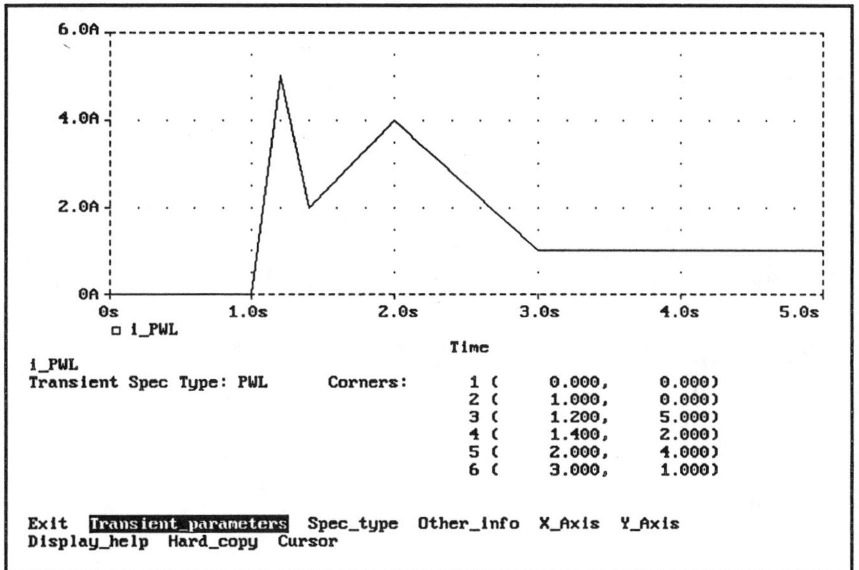

REPEAT ... ENDREPEAT loops are introduced to permit repetitions. They may appear anywhere a (<*tn*> <*in*>) pair appears. Currently they may not be nested. Absolute times within REPEAT loops are with respect to the start of the current iteration.

General Form

SFFM (<ioff> <iampl> <fc> <mod> <fm>)

Example

IMOD 10 5 SFFM(2 1 8Hz 4 1Hz)

Figure C-10: Independent Current Source and Stimulus Frequency-Modulated Waveform Parameters

Parameters	Description	Units	Default
<ioff>	Offset current	amp	none
<iampl>	Peak amplitude of current	amp	none
<fc>	Carrier frequency	hertz	1/TSTOP
<mod>	Modulation index		0
<fm>	Modulation frequency	hertz	1/TSTOP

Figure C-11: SFFM Current Waveform

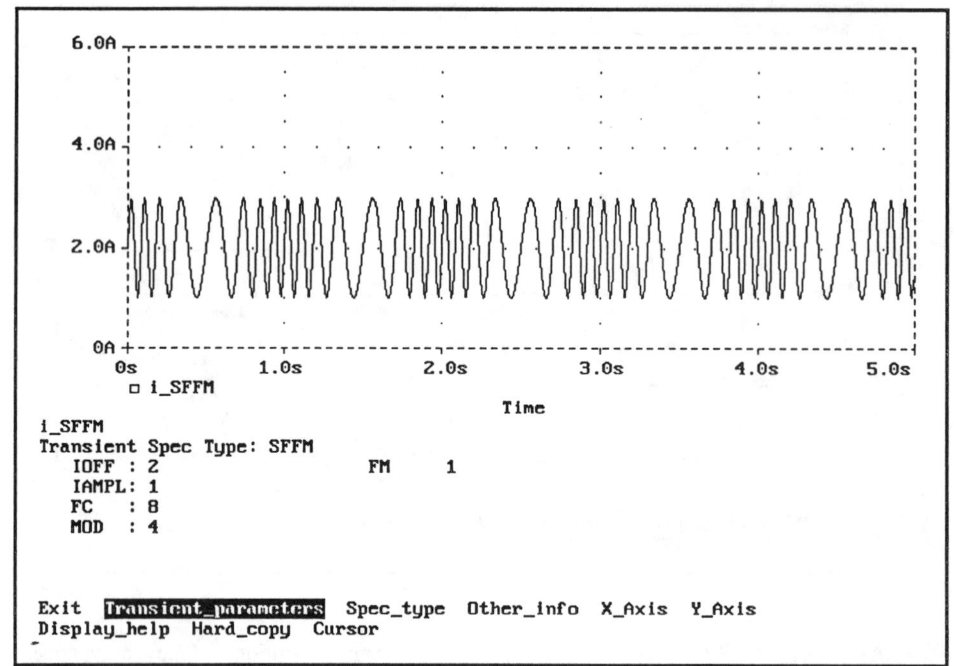

General Form

SIN (<ioff> <iampl> <freq> <td> <df> <phase>)

Example

ISIG 10 5 SIN(2 2 5Hz 1sec 10 30)

Figure C-12: Independent Current Source and Stimulus Sinusoidal Waveform Parameters

Parameters	Description	Units	Default
<ioff>	Offset current	amp	none
<iampl>	Peak amplitude of current	amp	none
<freq>	Frequency	hertz	1/TSTOP
<td>	Delay	sec	0
<df>	Damping factor	sec^{-1}	0
<phase>	Phase	degree	0

Figure C-13: SIN Current Waveform

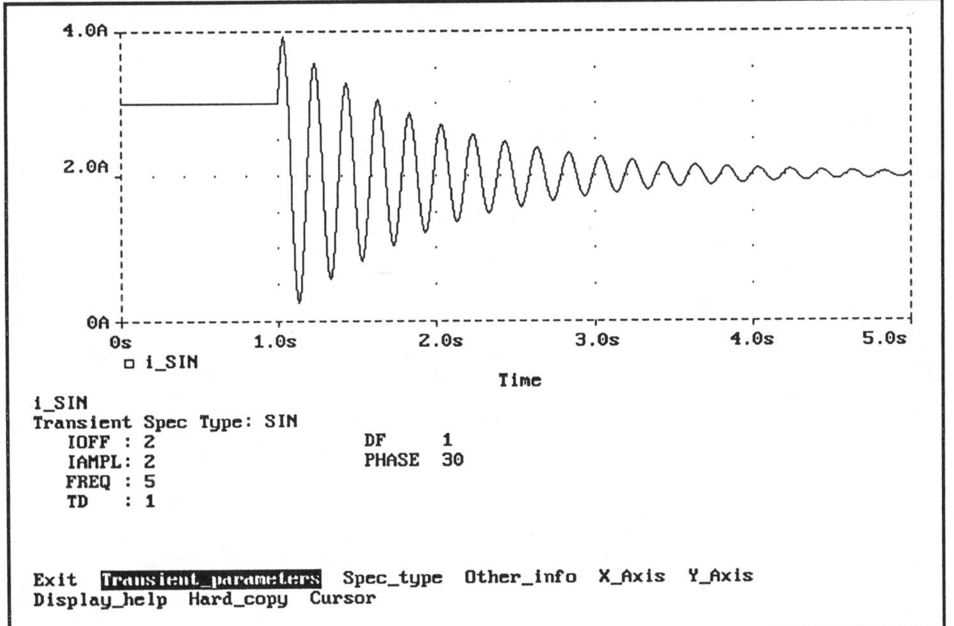

J — Junction FET

General Form

J<*name*> <*drain node*> <*gate node*> <*source node*>
+ <*model name*> [*area value*]

Examples

JIN 100 1 0 JFAST
J13 22 14 23 JNOM 2.0

Model Forms

.MODEL <*model name*> NJF [*model parameters*]
.MODEL <*model name*> PJF [*model parameters*]

Figure C-14: Junction FET Model Parameters

Model Parameters (see .MODEL statement)	Description	Units	Default
VTO	Threshold voltage	volt	-2.0
BETA	Transconductance coefficient	amp/volt^2	1E-4

Figure C-14: Junction FET Model Parameters (Continued)

Model Parameters (see .MODEL statement)	Description	Units	Default
LAMBDA	Channel-length modulation	volt^{-1}	0
IS	Gate *p-n* saturation current	amp	1E-14
N	Gate *p-n* emission coefficient		1
ISR	Gate *p-n* recombination current parameter	amp	0
NR	Emission coefficient for ISR		2
ALPHA	Ionization coefficient	volt^{-1}	0
VK	Ionization “knee” voltage	volt	0
RD	Drain ohmic resistance	ohm	0
RS	Source ohmic resistance	ohm	0
CGD	Zero-bias gate-drain *p-n* capacitance	farad	0
CGS	Zero-bias gate-source *p-n* capacitance	farad	0
M	Gate *p-n* grading coefficient		0.5
PB	Gate *p-n* potential	volt	1.0
FC	Forward-bias depletion capacitance coefficient		0.5
VTOTC	VTO temperature coefficient	volt/°C	0

K Inductor or Transmission Line Coupling

General Forms

K<*name*> L<*inductor name*> < L<*inductor name*> >*
+ <*coupling value*>

K<*name*> < L<*inductor name*> >* <*coupling value*>
+ <*model name*> [*size value*]

K<*name*>T<*transmission line name*>T<*transmission line name*>
+ Cm=<*capacitive coupling*> Lm=<*inductive coupling*>

Note: Transmission line coupling is not supported on the DOS low memory platform.

Examples

```
KTUNED L3OUT  L4IN .8
KTRNSFRM  LPRIMARY  LSECNDRY 1
KXFRM L1 L2  L3  L4 .98 KPOT_3C8
K2LINES  T1  T2  Lm=1m  Cm=.5p
```

Model Form

.MODEL <*model name*> CORE [*model parameters*]

Figure C-15: Inductor Coupling Model Parameters

Model Parameters (see .MODEL statement)	Description	Units	Default
LEVEL	Model index		1
AREA	Mean magnetic cross-section	cm^2	0.1
PATH	Mean magnetic path length	cm	1.0
GAP	Effective air-gap length	cm	0
PACK	Pack (stacking) factor		1.0
MS	Magnetization saturation	Gauss	1E+6
A	Thermal energy parameter	amp/meter	1E+3
C	Domain flexing parameter		0.2
K	Domain anisotropy parameter	amp/meter	500
ALPHA	Interdomain coupling parameter (**LEVEL**=1)		1e-3
GAMMA	Domain damping parameter (**LEVEL**=1)	sec^{-1}	infinite

Inductor Coupling

K<*name*> couples two, or more, inductors. Using the "dot" convention, place a "dot" on the first node of each inductor. In other words, given:

```
I1 1 0 AC 1mA
L1 1  0 10uH
L2 2  0 10uH
R2 2  0 .1
K12 L1 L2 1
```

the current through L2 will be in the opposite direction as the current through L1. The polarity is determined by the order of the nodes in the L device(s) and not by the order of inductors in the K statement.

<*coupling value*> is the "coefficient of mutual coupling" which must be between 0 and 1. This coefficient is defined by the equation

<*coupling value*> $= M_{ij}/(L_i \cdot L_j)^{1/2}$

where

L_i,L_j are a coupled-pair of inductors
M_{ij} is the mutual inductance between L_i and L_j

For transformers of normal geometry, the value 1 should be used. Values less than 1 occur in air core transformers when the coils do not completely overlap.

The linear branch relation for transient analysis is

$$V_i = L_i \cdot \frac{dI_i}{dt} + M_{ij} \cdot \frac{dI_j}{dt} + M_{ik} \cdot \frac{dI_k}{dt} + \cdots$$

L Inductor

General Form

L<*name*> <(+) *node*> <(-) *node*> [*model name*] <*value*>
+ [IC=<*initial value*>]

Examples

```
LLOAD 15 0 20mH
L2 1 2 .2E-6
LCHOKE 3 42 LMOD .03
LSENSE 5 12 2UH IC=2mA
```

Model Form

.MODEL <*model name*> IND [*model parameters*]

Figure C-16: Inductor Model Parameters

Model Parameters (see .MODEL statement)	Description	Units	Default
L	Inductance multiplier		1
IL1	Linear current coefficient	amp^{-1}	0
IL2	Quadratic current coefficient	amp^{-2}	0
TC1	Linear temperature coefficient	$°C^{-1}$	0
TC2	Quadratic temperature coefficient	$°C^{-2}$	0
T_MEASURED	Measured temperature	°C	
T_ABS	Absolute temperature	°C	
T_REL_GLOBAL	Relative to current temperature	°C	
T_REL_LOCAL	Relative to AKO model temperature	°C	

The (+) and (-) nodes define the polarity meant when the inductor has a positive voltage across it. Also, positive current flows from the (+) node through the inductor to the (-) node.

If [*model name*] is left out, then the effective value is <*value*>.

If [*model name*] is specified, then the effective value is given by the formula

<*value*>·**L**·(1+**IL1**·I+**IL2**·I^2)·(1+**TC1**·(T-Tnom)+**TC2**·$(T\text{-}Tnom)^2$)

where *<value>* is normally positive (though it can be negative, but **not** zero). "Tnom" is the nominal temperature (set with TNOM option).

If the inductor is associated with a Core model, then the effective value is the number of turns on the core. Otherwise, the effective value is the inductance.

<initial value> is the initial guess for the current through the inductor during the bias point calculation. See the "Standard Analysis; Setting Initial Conditions" section of the "Analysis Specification" chapter in the *Circuit Analysis User's Guide* for more information on setting initial conditions.

M MOSFET

General Form

M*<name>* *<drain node>* *<gate node>* *<source node>*
\+ *<bulk/substrate node>* *<model name>*
\+ [L=*<value>*] [W=*<value>*]
\+ [AD=*<value>*] [AS=*<value>*]
\+ [PD=*<value>*] [PS=*<value>*]
\+ [NRD=*<value>*] [NRS=*<value>*]
\+ [NRG=*<value>*] [NRB=*<value>*]
\+ [M=*<value>*]

Examples

```
M1 14 2 13 0 PNOM  L=25u W=12u
M13 15 3 0 0 PSTRONG
M16 17 3 0 0 PSTRONG M=2
M28 0 2 100 100 NWEAK L=33u W=12u
+ AD=288p AS=288p PD=60u PS=60u NRD=14 NRS=24 NRG=10
```

The following list describes the parameters common to all model levels, which are primarily parasitic element values such as series resistance, overlap and junction capacitance, and so on.

Figure C-17: MOSFET Model Parameters for All Levels

Model Parameters (see .MODEL statement)	Description	Units	Default
LEVEL	Model index		1
L	Channel length	meter	DEFL
W	Channel width	meter	DEFW
RD	Drain ohmic resistance	ohm	0
RS	Source ohmic resistance	ohm	0
RG	Gate ohmic resistance	ohm	0

Figure C-17: MOSFET Model Parameters for All Levels (Continued)

Model Parameters (see .MODEL statement)	Description	Units	Default
RB	Bulk ohmic resistance	ohm	0
RDS	Drain-source shunt resistance	ohm	infinite
RSH	Drain, source diffusion sheet resistance	ohm/square	0
IS	Bulk *p-n* saturation current	amp	1E-14
JS	Bulk *p-n* saturation current/area	amp/meter^2	0
JSSW	Bulk *p-n* saturation sidewall current/length	amp/meter	0
N	Bulk *p-n* emission coefficient		1
PB	Bulk *p-n* bottom potential	volt	0.8

Q — Bipolar Transistors

General Form

Q<*name*> < *collector node*> <*base node*> <*emitter node*>
+ [*substrate node*] <*model name*> [*area value*]

Examples

```
Q1 14 2 13 PNPNOM
Q13 15 3  0 1 NPNSTRONG 1.5
Q7 VC 5 12 [SUB] LATPNP
```

Model Forms

.MODEL <*model name*> NPN [*model parameters*]
.MODEL <*model name*> PNP [*model parameters*]
.MODEL <*model name*> LPNP [*model parameters*]

Figure C-18: Bipolar Transistor Model Parameters

Model Parameters (see .MODEL statement)	Description	Units	Default
IS	Transport saturation current	amp	1E-16
BF	Ideal maximum forward beta		100
NF	Forward current emission coefficient		1
VAF (VA)	Forward Early voltage	volt	infinite

Figure C-18: Bipolar Transistor Model Parameters (Continued)

Model Parameters (see .MODEL statement)	Description	Units	Default
IKF (IK)	Corner for forward-beta high-current roll-off	amp	infinite
ISE (C2)	Base-emitter leakage saturation current	amp	0
NE	Base-emitter leakage emission coefficient		1.5
BR	Ideal maximum reverse beta		1
NR	Reverse current emission coefficient		1
VAR (VB)	Reverse Early voltage	volt	infinite
IKR	Corner for reverse-beta high-current roll-off	amp	infinite
ISC (C4)	Base-collector leakage saturation current	amp	0
NC	Base-collector leakage emission coefficient		2
NK	High-current roll-off coefficient		.5
ISS	Substrate *p-n* saturation current	amp	0
NS	Substrate *p-n* emission coefficient		1
RE	Emitter ohmic resistance	ohm	0
RB	Zero-bias (maximum) base resistance	ohm	0
RBM	Minimum base resistance	ohm	RB
IRB	Current at which Rb falls halfway to	amp	infinite
RC	Collector ohmic resistance	ohm	0
CJE	Base-emitter zero-bias *p-n* capacitance	farad	0
VJE (PE)	Base-emitter built-in potential	volt	0.75
MJE (ME)	Base-emitter *p-n* grading factor		0.33
CJC	Base-collector zero-bias *p-n* capacitance	farad	0
VJC (PC)	Base-collector built-in potential	volt	0.75
MJC (MC)	Base-collector *p-n* grading factor		0.33
XCJC	Fraction of Cbc connected internal to Rb		1
CJS (CCS)	Substrate zero-bias *p-n* capacitance	farad	0
VJS (PS)	Substrate *p-n* built-in potential	volt	0.75
MJS (MS)	Substrate *p-n* grading factor		0
FC	Forward-bias depletion capacitor coefficient		0.5
TF	Ideal forward transit time	sec	0
XTF	Transit time bias dependence coefficient		0
VTF	Transit time dependency on Vbc	volt	infinite

Figure C-18: Bipolar Transistor Model Parameters (Continued)

Model Parameters (see .MODEL statement)	Description	Units	Default
ITF	Transit time dependency on Ic	amp	0
PTF	Excess phase @ 1/(2π·TF)Hz	degree	0
TR	Ideal reverse transit time	sec	0
QCO	Epitaxial region charge factor	coulomb	0
RCO	Epitaxial region resistance	ohm	0
VO	Carrier mobility "knee" voltage	volt	10
GAMMA	Epitaxial region doping factor		1E-11
EG	Bandgap voltage (barrier height)	eV	1.11

R Resistor

General Form

R<*name*> <(+) *node*> <(-) *node*> [*model name*] <*value*>
+ [TC = <TC1> [,<TC2>]]

Examples

```
RLOAD 15 0 2K
R2 1 2 2.4E4 TC=.015,-.003
RFDBCK 3 33 RMOD 10K
```

Model Form

.MODEL <*model name*> RES [*model parameters*]

Figure C-19: Resistor Model Parameters

Model Parameters (see .MODEL statement)	Description	Units	Default
R	Resistance multiplier		1
TC1	Linear temperature coefficient	$°C^{-1}$	0
TC2	Quadratic temperature coefficient	$°C^{-2}$	0
TCE	Exponential temperature coefficient	%/°C	0
T_MEASURED	Measured temperature	°C	

Figure C-19: Resistor Model Parameters (Continued)

Model Parameters (see .MODEL statement)	Description	Units	Default
T_ABS	Absolute temperature	°C	
T_REL_GLOBAL	Relative to current temperature	°C	
T_REL_LOCAL	Relative to AKO model temperature	°C	

For information on **T_MEASURED, T_ABS, T_REL_GLOBAL**, and **T_REL_LOCAL**, see the .MODEL statement on page 52.

The (+) and (-) nodes define the polarity meant when the resistor has a positive voltage across it. Positive current flows from the (+) node through the resistor to the (-) node.

Temperature coefficients for the resistor can be specified in-line, as in the second example. If the resistor **has a model specified,** then the coefficients from the model are used for the temperature updates, otherwise the in-line values are used. In both cases the temperature coefficients default to zero. Expressions **may not be used** for the in-line coefficients.

S Voltage-Controlled Switch

General Form

S<*name*> <(+) *switch node*> <(-) *switch node*>
+ <(+) *controlling node*> <(-) *controlling node*>
+ <*model name*>

Examples

```
S12    13 17  2 0 SMOD
SRESET  5  0 15 3 RELAY
```

Model Form

.MODEL <*model name*> VSWITCH [*model parameters*]

Figure C-20: Voltage-Controlled Switch Model Parameters

Model Parameters (see .MODEL statement)	Description	Units	Default
RON	"On" resistance	ohm	1.0
ROFF	"Off" resistance	ohm	1E+6
VON	Control voltage for "on" state	volt	1.0
VOFF	Control voltage for "off" state	volt	0.0

The voltage-controlled switch is a special kind of voltage-controlled resistor. The resistance between the <(+) *switch node*> and <(-) *switch node*> depends on the voltage between the <(+) *controlling node*> and <(-) *controlling node*>. The resistance varies continuously between **RON** and **ROFF**.

RON and **ROFF** must be greater than zero and less than 1/GMIN.

A resistance of 1/GMIN is connected between the controlling nodes to keep them from floating. See the .OPTIONS statement for setting GMIN.

We have chosen this model for a switch to try to minimize numerical problems. However, there are a few things to keep in mind:

> With double precision numbers, PSpice can only handle a dynamic range of about 12 decades. So, we do not recommend making the ratio of **ROFF** to **RON** greater than 1E+12.
>
> Similarly, we do not recommend making the transition region too narrow. Remember that in the transition region the switch has gain. The narrower the region, the higher the gain and the greater the potential for numerical problems. The smallest allowed value for **VON-VOFF** is**RELTOL·(MAX(VON ,VOFF))+VNTOL.**
>
> Although very little computer time is required to evaluate switches, during transient analysis PSpice must step through the transition region with a fine enough step size to get an accurate waveform. So, for many transitions you may have long run times from evaluating the other devices in the circuit many times.

V Independent Voltage Source and Stimulus

General Form

V<*name*> <(+) *node*> <(-) *node*>
+ [[DC] <*value*>]
+ [AC <*magnitude value*> [*phase value*]]
+ [*transient specification*]

Examples

```
VBIAS 13 0 2.3mV
VAC 2 3 AC .001
VACPHS 2 3 AC .001 90
VPULSE 1 0 PULSE(-1mV 1mV 2ns 2ns 2ns 50ns 100ns)
V3 26 77 DC .002 AC 1 SIN(.002 .002 1.5MEG)
```

This element is a voltage source. The default value is zero for the DC, AC, and transient values. None, any, or all of the DC, AC, and transient values may be specified. The AC phase value is in degrees.

Note: Positive current (as recorded in the output file, or for Probe, or as used as an input to a current-controlled source) is current flowing into the (+) node.

If present, [*transient specification*] must be one of:

EXP (<*parameters*>)	for an exponential waveform
PULSE (<*parameters*>)	for a pulse waveform
PWL (<*parameters*>)	for a piecewise linear waveform
SFFM (<*parameters*>)	for a frequency-modulated waveform
SIN (<*parameters*>)	for a sinusoidal waveform

The variables TSTEP and TSTOP, which are used in defaulting some waveform parameters, are set by the .TRAN command. TSTEP is *<print step value>* and TSTOP is *<final time value>*. The .TRAN command can be anywhere in the circuit file; it need not come after the voltage source.

General Form

EXP (<v1> <v2> <td1> <tc1> <td2> <tc2>)

Example

VRAMP 10 5 EXP(1V 5V 1 .2 2 .5)

Figure C-21: Independent Voltage Source and Stimulus Exponential Waveform Parameters

Parameters	Description	Units	Default
<v1>	Initial voltage	volt	none
<v2>	Peak voltage	volt	none
<td1>	Rise (fall) delay	sec	0
<tc1>	Rise (fall) time constant	sec	TSTEP
<td2>	Fall (rise) delay	sec	<td1>+TSTEP
<tc2>	Fall (rise) time constant	sec	TSTEP

Figure C-22: EXP Voltage Waveform

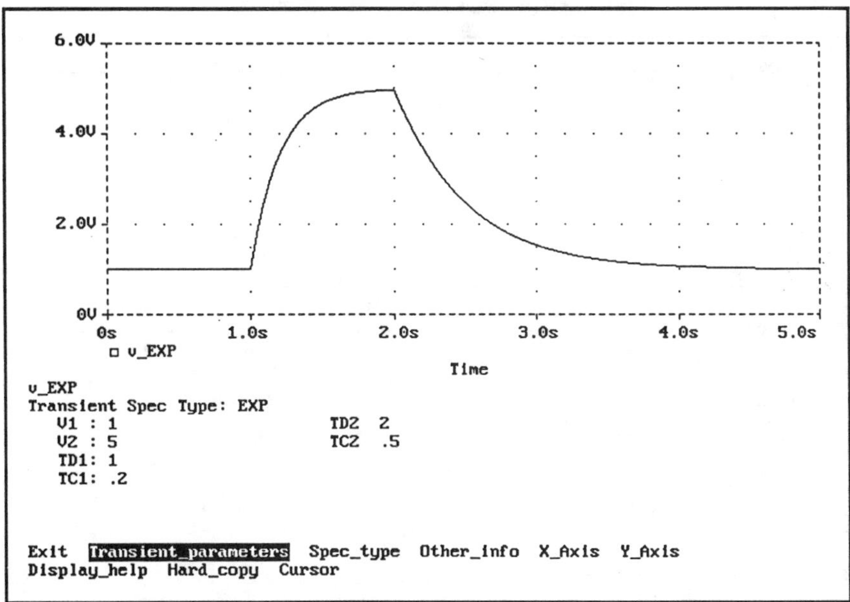

General Form

PULSE (<v1> <v2> <td> <tr> <tf> <pw> <per>)

Example

VSW 10 5 PULSE(1V 5V 1sec .1sec .4sec .5sec 2sec)

Figure C-23: Independent Voltage Source and Stimulus Pulse Waveform Parameters

Parameters	Description	Units	Default
<v1>	Initial voltage	volt	none
<v2>	Pulsed voltage	volt	none
<td>	Delay	sec	0
<tr>	Rise time	sec	TSTEP
<tf>	Fall time	sec	TSTEP
<pw>	Pulse width	sec	TSTOP
<per>	Period	sec	TSTOP

Figure C-24: PULSE Voltage Waveform

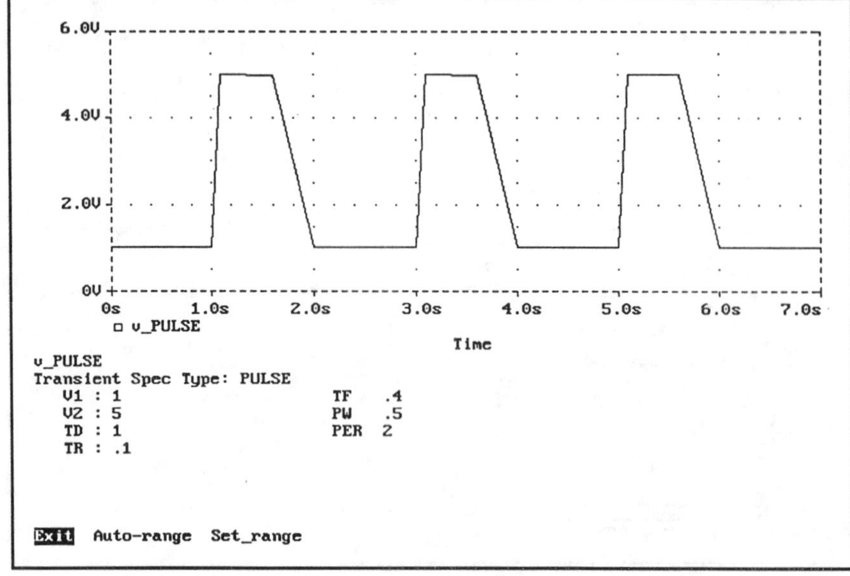

General Form

PWL [<*time_scale_factor*>] [<*value_scale_factor*>](<*tn*> <*vn*>)*
+ [[REPEAT FOR <*n*>] (<*tn*> <*vn*>)* [ENDREPEAT]]*
+ [[REPEAT FOREVER] (<*tn*> <*vn*>)* [ENDREPEAT]]*

or with time, value pairs (<*tn*>, <*vn*>) contained within a file:

PWL [<*time_scale_factor*>] [<*value_scale_factor*>] FILE<*file name*>

Examples

```
V3 10 5 PWL(O OV 1 OV 1.2 5V 1.4 2V 2 4V 3 1V)
```

N volt square wave (where N is 1, 2, 3, 4, then 5); 75% duty cycle; 10 cycles; 1 microseconds per cycle:

```
.PARAM N=1
.STEP PARAM N   1,5,1
V1 1 0 PWL
+     TIME_SCALE_FACTOR=1e-6   ; all time units are scaled to microseconds
+     REPEAT FOR 10
+       (0, 0)
+       (.25, 0)
+       (.26, {N})
+       (.99, {N})
+       (1, 0)
+      ENDREPEAT
```

5 volt square wave; 75% duty cycle; 10 cycles; 10 microseconds per cycle; followed by 50% duty cycle N volt square wave (where N is 1, 2, 3, 4, then 5) lasting until the end of simulation:

```
.PARAM N=.2
.STEP PARAM N   .2, 1.0, .2
V1 1 0 PWL
+     TIME_SCALE_FACTOR=1e-5   ; all time units are scaled to 10 microseconds
+     VALUE_SCALE_FACTOR=5
+     REPEAT FOR 10
+       (0, 0)
+       (.25, 0)
+       (.26, 1)
+       (.99, 1)
+       (1, 0)
+      ENDREPEAT
+     REPEAT FOREVER
+       (0, 0)
+       (+.50, 0)
+       (+.01, {N})         iteration time .51
+       (+.48, {N})         iteration time .99
+       (1, 0)
+       ENDREPEAT
```

The PWL form describes a piecewise linear waveform. Each pair of time-voltage values specifies a corner of the waveform. Up to 3995 pairs may be used for the IBM-PC and the Macintosh. The current at times between corners is the linear interpolation of the currents at the corners. This behavior is shown in Figure C-25.

Figure C-25: PWL Voltage Waveform

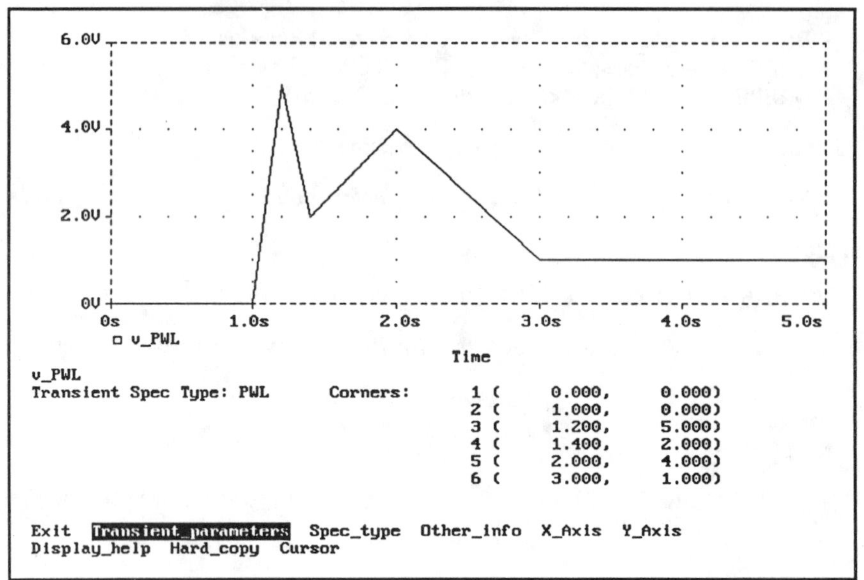

General Form

SFFM (<voff> <vampl> <fc> <mod> <fm>)

Example

vfm 10 5 sffm(2v 1v 8Hz 4 1Hz)

Figure C-26: Independent Voltage Source and Stimulus Frequency-Modulated Waveform Parameters

Parameters	Description	Units	Default
<mod>	Modulation index		0
<voff>	Offset voltage	volt	none
<fm>	Modulation frequency	hertz	1/TSTOP
<vampl>	Peak amplitude of voltage	volt	none
<fc>	Carrier frequency	hertz	1/TSTOP

Figure C-27: SFFM Voltage Waveform

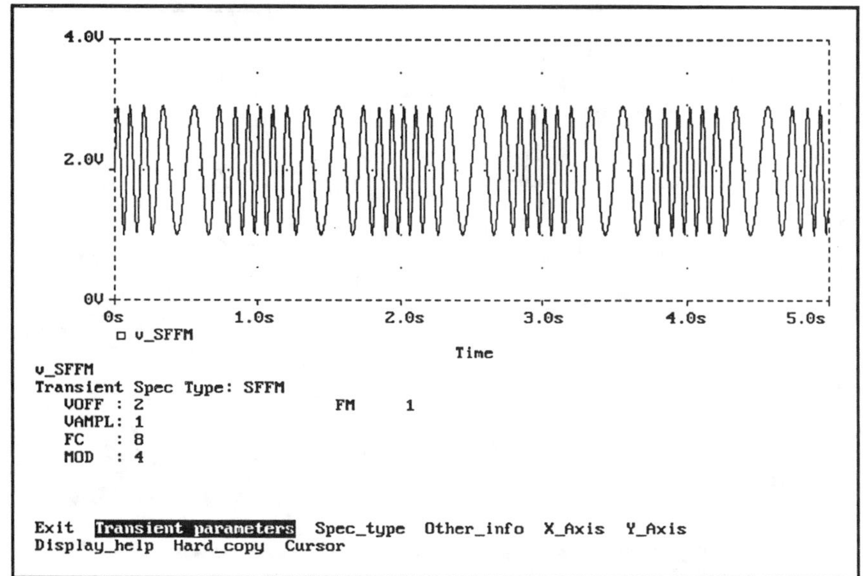

General Form

SIN (<voff> <vampl> <freq> <td> <df> <phase>)

Example

VSIG 10 5 SIN(2 2 5Hz 1 10 30)

Figure C-28: Independent Voltage Source and Stimulus Sinusoidal Wave-form Parameters

Parameters	Description	Units	Default
<voff>	Offset voltage	volt	none
<vampl>	Peak amplitude of voltage	volt	none
<freq>	Frequency	hertz	1/TSTOP
<td>	Delay	sec	0
<df>	Damping factor	sec^{-1}	0
<phase>	Phase	degree	0

Figure C-29: SIN Voltage Waveform

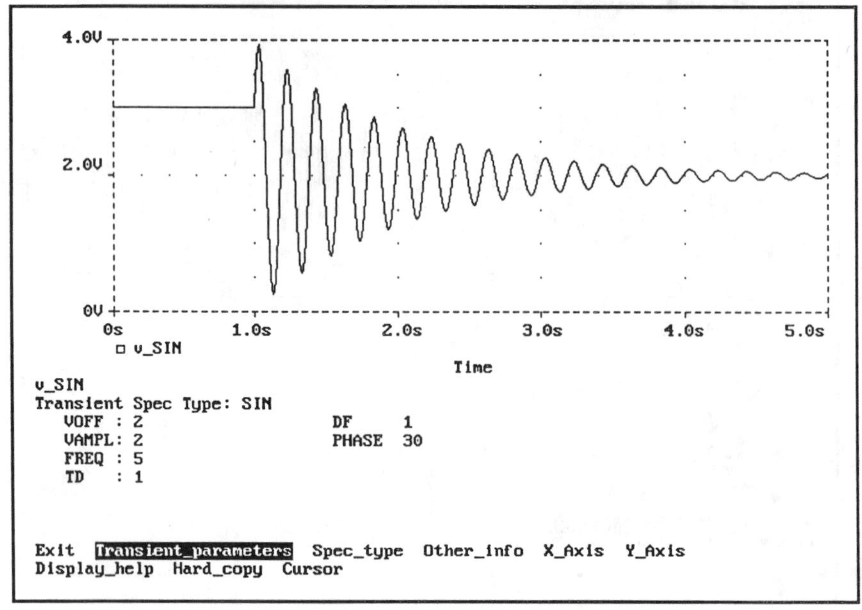

W Current-Controlled Switch

General Form

W<*name*> <(+) *switch node*> <(-) *switch node*>
+ <*controlling V device name*> <*model name*>

Examples

W12 13 17 VC WMOD
WRESET 5 0 VRESET RELAY

Model Form

.MODEL <*model name*> ISWITCH [*model parameters*]

Figure C-30: Current-Controlled Switch Model Parameters

Model Parameters (see .MODEL statement)	Description	Units	Default
RON	"On" resistance	ohm	1.0
ROFF	"Off" resistance	ohm	1E+6
ION	Control current for "on" state	amp	1E-3
IOFF	Control current for "off" state	amp	0.0

The current-controlled switch is a special kind of voltage-controlled resistor. The resistance between the <(+) *switch node* and <(-) *switch node*> depends on the current through <*controlling V device name*>. The resistance varies continuously between **RON** and **ROFF**.

RON and **ROFF** must be greater than zero and less than 1/GMIN.

A resistance of 1/GMIN is connected between the controlling nodes to keep them from floating. See the .OPTIONS statement for setting GMIN.

We have chosen this model for a switch to try to minimize numerical problems. However, there are a few things to keep in mind:

> With double precision numbers, PSpice can handle only a dynamic range of about 12 decades. Therefore, we do not recommend making the ratio of **ROFF** to **RON** greater than 1E+12.
>
> Similarly, we do not recommend making the transition region too narrow. Remember that in the transition region the switch has gain. The narrower the region, the higher the gain and the greater the potential for numerical problems. The smallest allowed value for **ION-IOFF** is **RELTOL·(MAX(ION ,IOFF))+ABSTOL.**

X Subcircuit Call

General Form

> X<*name*> [*node*]* <*subcircuit name*> [PARAMS: <<*name*> = <*value*>>*]
> + [TEXT: < <*name*> = <*text value*> >*]

Examples

```
X12 100 101 200 201 DIFFAMP
XBUFF 13 15 UNITAMP
```

<*subcircuit name*> is the name of the subcircuit's definition (see .SUBCKT statement). There must be the same number of nodes in the call as in the subcircuit's definition. This statement causes the referenced subcircuit to be inserted into the circuit with the given nodes replacing the argument nodes in the definition. It allows you to define a block of circuitry once and then use that block in several places.

The keyword PARAMS: allows values to be passed into subcircuits as arguments and used in expressions inside the subcircuit. See the "Device Declarations; Component Values as Expressions" section of the "Circuit File Construction" chapter in the *Circuit Analysis User's Guide* for more information on this capability.

The keyword TEXT: allows text values to be passed into subcircuits, and to be used in text expressions inside the subcircuit.

Subcircuit calls may be nested. That is, you may have a call to subcircuit A, whose definition contains a call to subcircuit B. The nesting may be to any level, but **must not be circular**: for example, if subcircuit A's definition contains a call to subcircuit B, then subcircuit B's definition must not contain a call to subcircuit A.

Index